Records of Achievement

Report of the National Evaluation of Pilot Schemes

Patricia Broadfoot
Mary James
Susan McMeeking
Desmond Nuttall
Barry Stierer

A report submitted to the Department of Education and Science and the Welsh Office by the Pilot Records of Achievement in Schools Evaluation (PRAISE) team.

© Crown copyright 1988
First published 1988
ISBN 0 11 270657 6

Contents

The Pilot Records of Achievement in Schools Evaluation (PRAISE) team was commissioned by the Department of Education and Science and the Welsh Office to evaluate progress and results of the nine records of achievement pilot schemes funded from Education Support Grant from April 1985 to March 1988. The PRAISE team was based jointly at the University of Bristol School of Education and the Open University School of Education.

The names, roles and special contributions of team members to the production of this report were as follows:

Bristol University

Patricia Broadfoot
Part-time co-director
— case study of one school
— across-scheme analysis of project directors' and local evaluation reports
— synthesis

Barry Stierer
Full-time research associate
— case studies of seven schools
— across-site analysis of school case studies

Sue Cotterell
Project secretary 1985–86

Sheila Taylor
Project secretary 1987–88

Open University

Desmond Nuttall
Part-time co-director
— meta-evaluation strategy
— across-scheme analysis of project directors' and local evaluation reports

Mary James
Full-time deputy director and research fellow
— case study methodology
— case studies of six schools and one tertiary college
— across-site analysis of school case studies

Sue McMeeking
Full-time research assistant
— case studies of seven schools

Phil Clift
Part-time team member
— 'read and react' exercise in lieu of teachers' survey (see Introduction)

Robert McCormick
Part-time team member
— initial analysis of a selection of project directors' reports
— principal critical reader of draft case studies

Colin Morgan
Part-time consultant
— case study of one school

June Evison
Project secretary

Diana Griffiths
Audio-typist

Other

Dave Ebbutt
Part-time consultant (Homerton Educational Research and Development Unit)
– LEA policy study

Notes:

1. Although the above list gives special responsibilities, all aspects of the evaluation were developed through regular team discussion. Drafts of reports were revised in the light of comments and criticisms from team colleagues and study participants.

2. The case study reports of individual institutions on which much of this report is based are published separately by Bristol University and the Open University Schools of Education. These are available from the Research Unit, Bristol University School of Education, 22 Berkeley Square, Bristol, BS8 1HP.

We acknowledge with gratitude and genuine respect the assistance and co-operation of the pupils and teachers in the 22 schools featured in the across-site analysis of case studies of schools, as well as the project directors and local evaluators upon whose reports our across-scheme analysis is based. We have depended in all our work upon their generosity and good will, and we sincerely hope that we have done justice to their efforts as pioneers in comparatively uncharted terrain.

We are also grateful for all the help and support given to us by members of the Records of Achievement National Steering Committee (RANSC), and in particular the Evaluation Sub-committee.

Finally, we would like to thank all those colleagues in schools, LEAs and universities who encouraged us by responding so positively to our interim reports.

Since the passing of the 1944 Education Act which heralded the provision of secondary education for all children, there has been a steadily increasing expectation that achievements during this phase of schooling should be marked by some form of certification. In the words of chapter 10 of the 1963 Newsom Report[1]: 'Boys and girls who stay at school until they are 16 may reasonably look for some record of achievement when they leave'. For many young people this expectation has been met by the award of examination certificates by external examining bodies. However, public examinations were never intended to cater for all young people, and some boys and girls still leave school with no record of their experiences and achievements during their years of compulsory schooling. When in 1973 the school-leaving age was eventually raised to 16, the acknowledged problem was further highlighted by the fact that those for whom public examinations were not designed would have to remain in school for an additional year. In this context it became especially important to work towards the creation of worthwhile goals which would give all young people a sense of achievement which might be appropriately recognised. This aspiration came to be widely shared and has resulted in a number of explicit attempts to provide youngsters with records of achievement, sometimes known as profiles.

During the 1970s and early 1980s, therefore, a number of independent initiatives were undertaken in this respect. Although they varied widely in approach, they had the common aim of seeking to provide school leavers of all levels of attainment with a positive statement of achievement across a range of activities. An increasingly coherent philosophy also emerged. This emphasised assessment as part of the educational process and made pupils equal partners with teachers by giving them a prominent and active role. It also emphasised a need for positive, constructive, detailed and useful records which would support learning in school and help the school leaver to provide employers, trainers and other potential 'users' with the kind of information they required. Above all, it emphasised that the processes of recording should develop the self-concept, self-confidence, self-esteem and motivation of young people in such a way that they could gain the maximum benefit from their education.

In November 1983 the Government demonstrated that it was aware of this trend by issuing a draft policy statement on 'records of achievement for all school leavers'. Comments received included suggestions and criticisms but indicated almost unanimous support for such records. In the autumn of 1984, therefore, the Department of Education and Science and the Welsh Office published *Records of Achievement: A Statement of Policy* which incorporated many of the suggestions on the draft and presented the case for records of achievement based on the philosophy outlined above. As such it was widely welcomed by education professionals and became a key reference point for those involved in developing recording systems.

According to the 1984 policy statement the purposes of records of achievement were seen to be:

● to recognise, acknowledge and give credit for what pupils have achieved and experienced, not just in terms of results in public examinations but in other ways as well.

● to contribute to pupils' personal development and progress by improving their motivation, providing encouragement and increasing their awareness of strengths, weaknesses and opportunities.

● to help schools identify the all round potential of their pupils and to consider how well their curriculum, teaching and organisation enable pupils to develop the general, practical and social skills which are to be recorded.

---

[1] Ministry of Education: Central Advisory Council for Education *Half Our Future* HMSO (1963).

● to provide young people leaving school or college with a short, summary document of record which is recognised and valued by employers and institutions of further and higher education.

In addition to outlining purposes and processes, the 1984 policy statement indicated the Government's intention to issue national guidelines by the end of the decade for the introduction of records of achievement for all pupils in secondary schools. The Secretaries of State for Education and Science and for Wales believed that the next step should be to gain more experience of records of achievement through pilot schemes. Their aim was to establish the greatest possible degree of agreement on the main issues, including those identified in the policy statement, namely:

● the purposes of records and the recording system;

● the target group;

● the treatment of personal achievements and qualities;

● the presentation of examination results and other evidence of educational attainment;

● national currency and accreditation;

● recording processes, including the relationship with existing reporting systems;

● ownership;

● the requirements for in-service training of teachers;

● financial implications.

Pilot schemes Under the Education (Grants and Awards) Act 1984, the Department of Education and Science, in partnership with local education authorities, provided financial support for nine pilot schemes involving total expenditure of some £2.25 million a year in England over three years, with the Welsh Office providing proportional support for pilot work in Wales.

The nine schemes chosen to conduct pilot work between April 1985 and March 1988 included both existing and new schemes. Six schemes were based in single LEAs (Dorset, Essex, ILEA, Lancashire, Suffolk and Wigan) and three were multi-authority consortia (East Midlands Records of Achievement Project (EMRAP) comprising Derbyshire, Lincolnshire, Northamptonshire and Notting-hamshire; Oxford Certificate of Educational Achievement (OCEA) comprising Coventry, Leicestershire, Oxfordshire and Somerset; and the Welsh scheme comprising all eight LEAs in Wales).

All pilot schemes were expected to report regularly to the Records of Achievement National Steering Committee (RANSC) which was set up by the Secretaries of State to oversee the monitoring and evaluation of pilot schemes. This committee was made up of educationists, industrialists and others and was chaired by an Assistant Secretary at the DES. In the autumn of 1988 it was expected to report on experience gained in the pilot schemes and to offer draft national guidelines for consideration by the Secretaries of State.

National evaluation In addition to pilot scheme directors' reports, evidence to assist the deliberations of RANSC was to be supplied by a national evaluation. Tenders were invited from research bodies and a team of researchers from both the University of Bristol and the Open University Schools of Education was contracted to evaluate the progress and results of the nine records of achievement pilot schemes from 1985 to 1988. Total funds available to the evaluation project were £239,000 over three years (at 1985 prices).

The national evaluation, which came to be known as PRAISE (Pilot Records of Achievement in Schools Evaluation), was initially designed as four distinct strands. The evaluation evolved during its lifetime, partly in response to the emerging evidence, and partly as a result of external influences such as teachers' union action during 1985–86. However, the original strands were:

Since national guidelines for records of achievement will have to be implemented in schools and classrooms, the PRAISE evaluators felt it appropriate to invest most time and effort in studying the development and implementation of recording systems in individual institutions. Twenty-three schools and one tertiary college were originally selected for intensive fieldwork (approximately 300 days in total) and case study reports were written and cleared for all but two of these. These case study reports were expected to fulfil several purposes: to provide a second order data-base for an across-site analysis as part of our final report to RANSC (Part One of this report); to provide a source of illumination and guidance for teachers and others who will be expected to implement national guidelines in the future; and to provide feedback to schools participating in the study.

1 Case Studies

Each of the nine pilot schemes appointed its own local evaluator to monitor progress within the scheme and report to the LEA or consortium. These local evaluators adopted a variety of styles depending on the nature of the scheme and the resources available for evaluation. The original intention of the national evaluators was to support, collate and carry out a meta-analysis of evaluation work carried out by local evaluators, placing that work in the context of the national evaluation and making it available to RANSC. To this end regular six-monthly meetings were held between national and local evaluators, which served as a forum to share ideas and experience.

2 Meta-evaluation

As time progressed it became increasingly evident that the original aspirations for a systematic meta-evaluation would not be fulfilled because local evaluators had such different objectives, methodologies, reporting styles, and timetables. Indeed in some instances reports were not forthcoming at all, or they could not be made available to us at a time that would fit in with our reporting schedule, or they did not contain the kind of **evidence** that would provide us with a sound basis from which to draw general conclusions at national level. However, at the same time as hope for a comprehensive meta-evaluation of local evaluations was beginning to fade, a need emerged for an analysis of the reports from project directors that were submitted annually to RANSC. (In some cases the project director and the local evaluator were one and the same person.) For this reason the PRAISE Team decided to develop this strand of its work in terms of an across-scheme analysis of available reports from both project directors and local evaluators (Part Two of this report).

In the LEA policy study, the PRAISE Team hoped to study the efforts of local authorities to formulate and implement records of achievement policies within schools and colleges. Issues of likely importance included the links between national and local policy, links between records of achievement policies and other educational policies, and implications for organisational structures, implementation and dissemination strategies, INSET and resources.

3 LEA Policy Study

Work on this strand was delayed by a number of internal and external difficulties. However, in July 1987 a meeting of senior LEA officers and advisers was convened to identify and explore some of the policy issues and strategies emerging at local level. A paper derived from this experience was then used as a basis for further discussions in Oxfordshire and the ILEA and a case study of Wigan LEA. Since this strand of our evaluation was very much more limited than either of the two described above, and consistently confirmed the project directors' insights recorded in their reports, the results do not have a separate section in our report but were used to inform the synthesis in Part Three.

Initially, it was hoped that this strand of our evaluation would also incorporate a 'user context study' which would explore both the involvement of employers, parents and others in the formulation of policy or in local accreditation processes, and the responses of 'users' of summative documents. For the most part, however, these aspects were picked up within the case studies of schools and/or within project directors' reports, and there appeared to be no strong reason for conducting a separate study. Indeed, it proved exceedingly difficult to collect data on the response of users to summative documents because few schools had issued such documents by the time our data collection ended. The separately-funded study of user response, currently being carried out by the NFER and due to report in 1989, is likely to be a more useful source of information in this respect, because it is studying mainly summative records issued to pupils in 1988.

4 The Teachers' Survey   The fourth strand of our work was intended to collect evidence of the general impact of records of achievement on the work and attitudes of teachers by means of a questionnaire distributed to approximately 5000 teachers in 100 project schools (50 ESG pilot schools and 50 non-ESG project schools). It was hoped that this would test some of the hypotheses emerging from our case study work although it was acknowledged that interpretation of results might prove difficult if, as was likely, responses were context-specific. The survey was postponed because of the teachers' union action, and by the time it could be carried out our own interim findings were confirming the context-specific nature of teachers' attitudes. The Evaluation Sub-committee of RANSC therefore advised us to abandon the idea of a teachers' survey in favour of an exercise in which a sample of schools were asked to read and react to our 1987 interim report, with special reference to the extent to which our findings were consistent with their experience. Thus a sample was drawn up of 60 schools from the nine ESG-funded schemes and the nine non-ESG schemes. Responses were fed into the relevant sections of our final report insofar as they confirmed or disconfirmed the themes and issues we had identified, or raised others which we had not hitherto considered. In general our analysis received broad support from these schools and, as with the LEA policy study, it was not felt necessary to devote a separate section of our final report to this strand of our work.

As this brief description indicates, the four strands of our evaluation work were never expected to be equivalent in scope or depth and some strands assumed much greater importance than others as the pilot period progressed. Thus the case studies of schools and the analysis of project directors' and local evaluation reports became the major elements, whilst the teachers' survey and the LEA policy study were very limited activities whose major role was in the cross-validation of insights emerging in the major strands.

This report   Within all four strands of our evaluation, data were analysed in relation to the objectives and issues identified in the 1984 statement of policy and in the light of issues and concerns that emerged locally and nationally during the pilot period. Interim findings were presented in reports made available in 1986 and 1987. The 1987 interim report[1] was a substantial document which influenced the content and structure of this final report. Indeed, the themes and issues in this report are presented in much the same way as they were in our 1987 interim report although we now have more confidence in our analysis and the evidence we have used to support it. This has enabled us to move from identification of issues towards judgement, although we acknowledge that issues remain unresolved in some areas.

In relation to our analysis we feel that we should make three points quite clear at this stage. First, our analysis and conclusions pertain only to pilot work within the

---

[1] *PRAISE: Interim Evaluation Report* Open University School of Education and Bristol University School of Education, Milton Keynes and Bristol (mimeograph)(1987).

ESG-funded pilot schemes. Although individual team members have some knowledge of developments elsewhere, either through the Consultative Committee set up by the DES or through networks or informal contacts, the PRAISE team was solely contracted to evaluate progress within the DES pilot schemes. Readers may wish to consider the implications of this when studying our report because we make no claim to tell the whole story of records of achievement development at this time. Most importantly we have not been able to assess what is, or is not, possible without the provision of ESG funds. For this reason, whilst we make judgements based on our evidence, we make no hard and fast predictions about what will happen, or recommendations about what should happen, in the post-pilot phase. We feel that the latter is the proper function of RANSC when it has had an opportunity to consider all the evidence available to it, including the report of a similar study by HMI and some evidence from non-ESG initiatives.

Secondly, we would wish to emphasise that our account of progress in the pilot schemes has already acquired an historical quality. Data were mostly collected in the period from the end of 1985 to the end of 1987 and we know from our subsequent conversations with project co-ordinators in schools and schemes that much development has taken place in the interval between the close of data-collection and the publication of this report. This was inevitable but readers are asked to bear this in mind, especially when reading the illustrative material from schools in Part One.

Thirdly, we decided to adopt a thematic and issues-focused approach to our analysis on the grounds that this is the best way of providing relevant information for policy-makers, at national, local or school level. However, we recognise that this has led us away from making holistic judgements on the success, or otherwise, of the records of achievement initiative in general, or the pilot schemes in particular. In order to redress this imbalance somewhat, we would like to take this opportunity to say that whatever the particular problems that have faced schools and schemes, we have been consistently impressed by the enthusiasm and hard work of all those who have had a hand in development and implementation, sometimes in exceedingly difficult circumstances.

Report structure

The main body of our report is divided into three parts which reflect the main elements in our data-collection and analysis.

Part One is an across-site analysis based on the case study strand of our work, supported by the 'read and react' exercise which took the place of the teachers' survey. This provides a commentary on the major substantive themes and issues arising from the introduction of records of achievement in case study schools as well as associated management considerations. This commentary is supported with evidence selected from our 22 case study reports.

Part Two provides an interpretative overview of progress reports from the directors of the nine DES/Welsh Office records of achievement pilot schemes as well as the reports from local evaluators where these have been available. This part presents the results of what we originally conceived as the meta-evaluation strand of our work.

These two parts are substantially different in terms of the data sources upon which they are based, in that the first presents and analyses our own research data and the second reviews the reports emanating from the local schemes. In this sense the first part is based on primary sources, whilst the second is based on secondary sources. Moreover, Part One addresses school-level issues whereas Part Two conveys the perspective of those with LEA or scheme-wide responsibilities and concerns. Together these two parts highlight the very wide range of issues which are addressed in the debate on the development and implementation of records of achievement, and we hope they will be read in conjunction with each other.

Finally, in **Part Three** we have drawn together the insights offered in the first two parts of our report and in the small LEA policy study, and attempted to evaluate these against a framework of criteria which might be taken to constitute 'success' in recording achievement. On another level we have also tried to distil the major practical lessons learned from the experience of pilot schools and schemes in order to provide some guidance to all those who will be expected to respond to the need to develop records of achievement after the publication of national guidelines.

Throughout our report we have been concerned to illustrate the wide variety of experience **within** schemes and **between** schools. We did not perceive schemes to be monolithic so we made no attempt to compare and contrast schemes, or indeed schools, in order to identify a 'best buy' in terms of a records of achievement system. We did not have the resources to collect all the evidence that would be necessary to make such a judgement. Moreover, we considered it unlikely that any one school or scheme would embody an ideal model, or, conversely, that other schools and schemes would have nothing worthy of serious consideration by policy-makers. In the light of experience we feel that our decision was the right one and although **we do not offer any comprehensive models of good practice** we believe that our accounts illuminate the range of possibilities on most of the relevant dimensions.

Extension work   Towards the end of 1987, the Secretaries of State decided to support an extension of grant for further development work on a narrow range of specific issues embracing the relationship of records of achievement with other comparable initiatives (TVEI, GCSE, LAPP etc.); records of achievement for 16 to 19 year olds in schools and use in post-19 institutions; relationship to FE or YTS profiles; IT/computer support; recording through the full age-range of the school; resource implications of setting up and maintaining recording systems; and dissemination to a wider range of schools within an authority. As a result of further bids, seven local authority schemes in England (Dorset, Coventry, Essex, ILEA, Northamptonshire, Oxfordshire and Wigan) and the Welsh consortium of the eight LEAs in Wales have been offered a total grant not exceeding £2.4 million over two years to March 1990. Likewise, PRAISE has been offered further funding (£108,000 in total) over the same period to evaluate and report progress in relation to the issues identified. Thus a further report may be expected in the summer of 1990.

PART ONE

Across-site analysis of case studies of schools

by Mary James and Barry Stierer

Contents

Contents

This first main part of our report is an across-site analysis of data collected during case study work in 21 schools and one tertiary college from the beginning of the autumn term 1985 to the end of autumn term 1987. Approximately 10 to 15 visits were made throughout this period to each of the pilot institutions in our study. During these visits, case study workers carried out interviews with a range of staff involved with the project, as well as with pupils; they observed a wide range of classroom activities, including teacher-pupil discussion; and they sat in on meetings of various kinds connected with the project's development at institutional level. Questionnaires were also sent to parents and other users of summative records where these could be traced. This across-site analysis was prepared by two members of the PRAISE national evaluation team but it draws on, and incorporates, evidence from the 22 case study reports of institutions written by the five fieldworkers. (Please see Appendix 1 for the introductory sections of each of the 22 institutional case study reports, and Appendix 2 for a more detailed description of methodology.)

The framework for our analysis was derived from the themes and issues identified in the 1984 DES policy statement on records of achievement, but it was extended to take into consideration a number of additional themes and issues which emerged in the light of experience as pilot work progressed. In other words, some of the questions we address were anticipated whilst others arose out of the data.

The analysis is presented in two main sections: the first concerns substantive issues associated with records of achievement in case study institutions, whilst the second focuses on the management of RoA systems within institutions. Within each main section, detailed sub-sections provide both identification and discussion of key topics. Each of these sections and sub-sections contains both commentary and evidence.

In the **commentary** we have attempted to clarify the nature of each topic. This 'mapping' exercise has often required us to outline the range of approaches we have observed and to set out the different ways in which relevant terms are understood. Thus we have attempted to describe, categorise and interpret the experience of case study institutions. However, in contrast to the interim evaluation report which we prepared in 1987, we now feel that we are in a position to come to a judgement on some of the issues which we have identified. Where this is the case we have indicated the nature of our comments in the text, as we have where we feel that an issue or question is unresolved and requires further consideration.

By way of **evidence** we provide a number of relevant extracts from case study reports to illustrate the diversity of issues and approaches which characterise each topic. In an attempt to make reading easier, these examples are boxed in colour in the text. No attempt has been made to detail every institution's position in relation to each heading, as this would have been laborious, unnecessary and impractical. Neither have we seen it as important to ensure that all institutions have an equal number of references. Nevertheless, we have been as scrupulous as possible in selecting evidence under each heading in order to reflect fairly the range of work we have studied. We must emphasise, however, that although our case study institutions represent a great variety of settings and approaches, **they cannot, and should not, be taken to be a systematically representative sample within and across all nine pilot schemes.** The purpose of the illustrations is rather to illuminate and clarify the general points at issue in the design and implementation of RoA systems at institutional level. For this reason we have not chosen extracts which 'typify' the project in each school. Readers are advised to refer to Appendix 1 for brief introductions to the projects in each school.

Records of Achievement

While preparing this analysis we were conscious of the fact that we were addressing a wide range of audiences, each with its own priorities and responsibilities. Our most immediate reporting responsibility was to the Records of Achievement National Steering Committee (RANSC). We wished, accordingly, to relate the evidence from our case studies to RANSC's national policy deliberations. However, we also had in mind LEA officers and advisers, RoA project directors and support staff, and examination board personnel, who are concerned with the development, implementation and co-ordination of RoA schemes, and for the formulation of local RoA policies. We were aware, too, that providers of INSET might benefit from the availability of school-level evidence and analysis. Finally, we wanted our report to be relevant to staff in schools, whom we know to be eager not only for information about RoA developments outside their own areas but also for source material on which to focus school-based in-service work and planning discussions.

In attempting to relate our evidence to the purposes of such diverse audiences, we have commented upon issues which range from the practical to the philosophical. Each of our audiences will want to give urgent consideration to some of these questions, and will treat others as issues for longer-term rumination. We recognise that a number of the issues we raise have no easy or immediate solutions, and we are aware that there are difficulties associated with acting upon some of the judgements we make. Nevertheless, we have chosen to go beyond the purely descriptive, to raise issues and make judgements, not only because it was expected of us, but also to underline our view that the introduction of records of achievement carries with it profound educational and managerial implications which should not be minimised or overlooked. We do not mean to suggest, however, that such questions must be resolved before work may legitimately begin.

The distribution of the 22 institutions featuring in this report, across the nine DES/Welsh Office-funded pilot schemes, is as follows:

Dorset	1
Essex	2
ILEA	1
Lancashire	2
Suffolk	2
Wigan	2
EMRAP (Derbys, Lincs, Northants, Notts)	4
OCEA (Coventry, Leics, Oxon, Somerset)	4
Wales	4

In each case, individual institutions have been given a pseudonym to help protect their anonymity. However, the pilot schemes in which the schools and colleges are taking part have been identified by name, in keeping with the ethical guidelines agreed at the start of the evaluation.

We have referred to the institutions collectively as 'schools' throughout, even though one of them is a tertiary college. Where that college is cited specifically in the text it is referred to as a tertiary college.

In this first section we examine the kinds of substantive achievements and experiences which the various recording processes and documents have attempted to cover. We have tried as far as possible to discuss the **content** of RoAs independently of the recording **methods** which have been used. (Recording processes are the specific focus of section 1.2.)

Here we first look at the broad picture and attempt to clarify our frame of reference. Next we turn our attention to specific aspects of achievement and examine the content of recording in the areas of (a) extra-curricular and out-of-school activities, (b) subject-specific achievement, (c) cross-curricular achievement and (d) personal and social qualities and skills. Finally, we give an overview.

The DES policy statement made the point that: 'the internal processes of reporting, recording and discussion between teacher and pupil should cover a pupil's progress and activities across the whole educational programme of the school, both in the classroom and outside, and possibly activities outside the school as well' (paragraph 16).

This implies that RoAs should be **comprehensive** in terms of coverage. However the meaning of 'the whole programme of the school', and 'a pupil's progress' needs to be examined. Indeed, coverage as applied to records of achievement can be regarded as having two interrelated dimensions: on the one hand it involves conceptions of the **curriculum**; on the other hand it involves conceptions of the **person**. With regard to the first, the whole educational programme can be interpreted narrowly with reference only to the formal curriculum (academic and/or pastoral elements) or it can be interpreted widely to include extra-curricular activities (within, and even outside, experience provided by the school). With reference to the person, progress can be interpreted as referring to actual achievements (what the pupil can do or has done), or it can refer to what the pupil has experienced, or even to qualities or attributes that the pupil possesses or has developed, or any combination of these. In subsequent sections we look at these various aspects of achievement separately; here we address more general questions about the range and nature of achievements recorded.

We are able to report that most of the schools we have studied aimed in their RoA systems to portray the 'whole person'. In many cases this objective has evolved, in the light of experience, from an initial position where objectives for development were more limited. Thus, a broad consensus has emerged within our case study schools that RoAs should aim for some form of comprehensive coverage. The evident diversity among these schools has therefore lain less in ultimate purpose than in the vantage points from which they have attempted to gain a comprehensive view of the whole person, and in the conceptions of achievement which comprise this comprehensive view. Factors which have accounted for this diversity include the ethos of the school, the school's forms of curricular organisation, the school's conception of comprehensiveness, and the starting point for the school's RoA project.

For example, in some schools with a well-developed and extensive pastoral curriculum, the aim was to enable the portrayal of the whole person from the perspective of that programme.

1.1
Recording achievement: what is being covered?

1.1.1
Range of coverage

Subject-specific achievement was covered by some of the work carried out by form tutors piloting the OCEA P Component at Wisteria School. For example, in the first year of the pilot form tutors on the pilot tutorial team devised a procedure which enabled pupils to save photocopies of pieces of good work in their folders. Pro-formas were produced which had spaces for comments, about the chosen piece of work and about general progress within the subject, by the pupil, the subject teacher, the form tutor and a 'parent or friend'. These were called PRAISE sheets, and they were attached to the piece of work which pupils selected for their folders. PRAISE sheets were envisaged to serve a variety of formative functions, including greater dialogue between pupil and subject teacher, as well as between subject teacher and form tutor, and a wider evidential base upon which interviews could be conducted and end-of-year statements prepared.

Pupils also prepared 'report books' in their first year during tutorial lessons, in which they reviewed their progress and attitudes in relation to each of their subjects.

In other schools, where tutorial systems were limited to a largely administrative function, and the emphasis was on traditional academic curricula, the portrayal of the the whole person, including aspects of personal achievement, was incorporated in the recording within the academic curriculum organisation.

Unit Credits at Hydrangea School, ILEA, were (usually) single sheets upon which the scope and objectives of a course Unit were briefly described and upon which a pupil's achievement in relation to those objectives was recorded. On each Unit Credit, subject departments were expected to include objectives which were derived from all four of the 'aspects of achievement' posited in the report by the Hargreaves Committee, *Improving Secondary Schools*, i.e. propositional knowledge, practical and applied knowledge, personal and social skills, and positive attitudes associated with motivation and commitment. The definitions of aspects of achievement III and IV overlap considerably with the kinds of 'personal qualities' and 'personal and social skills' which were explicit features of RoA systems in other institutions. However, whereas these latter notions of achievement featured mainly in tutorial-based recording systems, attempts were made by means of Unit Credits at Hydrangea to record pupils' achievement in the areas defined by aspects III and IV in the context of subject-based learning.

But the majority of schools attempted to achieve coverage of most aspects of an individual's achievement by recording within both the academic and tutorial organisation.

From the beginning of the pilot the project team at Rose School, EMRAP, with no previous experience of profiling, expressed a desire to develop a RoA system which would cover subject-specific achievement, cross-curricular achievement, personal qualities and extra-curricular and out-of-school achievements. At the close of the pilot phase, most subject departments had taken on board some RoA principles and processes in their work with first and second year pupils, and some form of recording of personal qualities and achievements had been incorporated into tutorial work with years one to five. The development of a specific cross-curricular element remained elusive however.

There were, however, a very small number of schools where comprehensive coverage of the curriculum and portrayal of the person in terms of the widest definition of achievement was not considered a priority. In theses cases the recording system focused primarily on the recording of personal qualities and achievements and was located solely in the pastoral curriculum.

Although some references to experience and achievement in the academic curriculum at Campion Upper School, Suffolk, appeared in pupil accounts of

'Experiences of school and work', and were discussed with form tutors, recording across the curriculum was never considered to be a central feature of the RPA. At scheme level the possibility of adding subject profiles (e.g. science), graded test results (e.g. modern languages) and a cross-curricular skills profile to the existing RPA, which concentrated on interests and personal qualities, was considered during the pilot phase. Academic profiling and curricular provision for the development of cross-curricular skills was also put onto the agenda for discussion at school level. However, there was no expectation that personal aspects of achievement should be combined with academic or skills components in a comprehensive record of achievement during the period of study. Although this was a long-term aspiration of some senior staff, most teachers and pupils associated the pilot initiative solely with the existing Suffolk RPA project and thus with the pastoral curriculum and personal recording.

With these few exceptions, our evidence points to general and steady progress towards the creation of RoA systems, at school level, which aim to provide an holistic view of the individual based on recording of achievement across the whole curriculum and outside it as well.

In the next four sub-sections we look more closely at the various ways in which achievement has been conceptualised, which in turn has had a bearing on the way schools have gone about developing and implementing RoA systems. Our aim is to describe and analyse experience in order for readers to consider the wide range of alternative approaches. At this stage it is difficult to make any firm judgements about which approach is to be preferred, although we make some comments about approaches which would seem to us to have the greatest long-term potential. The fact that only one of our case study schools claimed to have reached steady-state in this respect (although in the light of recent developments in the LEA even this was doubtful) suggests that considerable time needs to be allowed for the development of such frameworks for recording.

1.1.2
Aspects of achievement

Having given a general indication of the range of achievement which RoA systems attempted to cover, we now turn to the aspects of achievement comprising that range. For the purposes of our analysis we have found it convenient to think of this range in terms of four aspects. We should stress however that we have chosen these categories because they are in common usage; as we shall argue later they are not necessarily theoretically or empirically distinct either within an individual or within a school's recording system. However, they have been adopted here to enable us to describe an essentially inter-dependent and multi-dimensional initiative using the linear medium of writing.

(a) Extra-curricular and out-of-school interests and activities

All case study schools have aimed to cover extra-curricular and out-of-school interests and activities in their recording systems. At first view there seems much value in this since choice of leisure time activities and work experience often says much about an individual which is of potential interest to employers and can provide interesting material for discussion at interview (e.g. the Hawthorn RC School girl who runs a pig and rabbit breeding business and wants to make a career in agriculture). It also appears to have value as a way of strengthening pupil-tutor relationships by providing an opportunity for tutors to discover a different side of their pupils and by giving in-school achievement a new perspective.

We have evidence that authentication has been sought within some RoA systems – for example, confirming signatures on the personal record or signatures on verification slips.

Records of Achievement

Often, however, what pupils say they have done has been taken on trust. Teachers appear reluctant to insist on written testimony from other adults precisely because such an act would imply suspicion and therefore undermine the developing relationship. Teachers also maintained that most pupils are basically honest.

In relation to this aspect of achievement we would wish to make four points. First, placing extra-curricular and out-of-school achievement on the recording agenda may imply an increased responsibility on the part of the school to provide extra-curricular activities. Secondly, teachers need to be acutely aware of the possibilities for social divisiveness and personal intrusion inherent in such an aspect of recording. Thirdly, pupils need to be made aware of the possible uses to which their records might be put to enable them to be both wide-ranging and judicious in their choice of evidence to record. Fourthly, pupil autonomy is an important principle to protect in the context of this aspect of recording.

(b) Subject-specific achievement

With the exception of two schools whose RoA was confined to personal achievement, all of the schools we have studied have, at some point during the pilot period, attempted to include pupils' achievement in their academic subjects within the RoA system. Hence we are able to report an emerging consensus in this area, as we were in the area of extra-curricular and out-of-school activities in the previous sub-section.

However, the scale and nature of such recording varied enormously according to the vantage points from which the recording of subject-specific achievement was carried out, and according to the notions of subject-specific achievement which underpinned it.

This diversity can be represented by describing **three broad types of recording of subject-specific achievement.** These are not 'ideal' types in any pure sense, and aspects of more than one type may be found within a single school. Nevertheless, we have found them to be a helpful way to understand the range of practices and definitions which abound in this area.

Within the **first type**, pupils' achievement in their subjects was recorded from a 'tutorial' vantage point. This type was generally built upon the responsibilities of form tutors for keeping an overall 'watching brief' on pupils' academic progress. Hence, whilst this vantage point held the potential for a comprehensive view of a child's experience of the academic curriculum, in the cases we studied it tended to be generalised and partial, lacking the specificity of assessment concepts relevant to each subject or course area. Often such records had the familiar character of the form tutor's overview found in conventional reports to parents or school leavers' testimonials. Certainly this type did not in itself require changes in the way that individual subject departments conceptualised achievement in their areas or in the way they organised teaching, learning, assessment and recording.

their progress in specific subjects in their 'Think Books' and to review such progress in one-to-one interviews with their tutor, and to incorporate such reviews in their end-of-year statements.

Within the **second type**, pupil's achievement in their subjects was recorded by subject teachers, but the emphasis of the RoA system was on **summarising** achievement and **reporting** that portrayal to particular outside audiences at certain times in a pupil's school career (we are not referring here exclusively to summative RoAs). Reporting frameworks were generally devised to reflect and incorporate existing methods of teaching, learning and assessment, and internal recording. For example, course work would be taught and marked by teachers in a traditional way, often using marks or grades, but achievement over a given period would be expressed in the form of positive prose comments, and discussed with pupils, at the point when summary reports were required.

There was never much sympathy at Magnolia Girls Grammar School, EMRAP, for the idea that new systems of teaching, learning and assessment should be instituted as part of the RoA initiative. Therefore the bulk of the development effort went into considering what should be **reported** and how. A record was made of subjects studied and examinations entered during each girl's career at the school, although no real value was perceived in giving details of topics and skills covered within syllabuses. (Views solicited by the co-ordinator from employers indicated that detailed accounts of work covered in each separate subject were mostly not considered important.) In addition an academic record was proposed which was intended to cover all subjects. This was seen initially as an edited version of subject reports to parents. As an interim measure a two-tier system of reports was suggested. Full subject reports to parents were expected to contain internal examination results, predictions of future success and assessments of weaknesses, as they have always done, but these elements were excised for the summative RoA in which only comments which were positive and non-predictive were to be included. However, considerable effort was directed at improving the quality of reports overall, in particular in encouraging staff to be specific about achievements and providing evidence where possible. The school saw little point in developing anything more elaborate than this, particularly in the light of direction from the LEA project team that summative documents should be no more than four sides of A4 (although additional evidence could be included in the folder). More detailed records would have to be drastically reduced and staff were not convinced that the result would be any better than the approach using conventional assessments and school reports which they proposed.

Within the **third type**, recording grew out of a re-appraisal by subject staff of notions of subject-specific achievement and a re-structuring of teaching, assessment and record-keeping. The emphasis here was on such formative processes as the explication of learning objectives, the development of modules or units of work, continuous criterion-referenced assessment, the grounding of records in evidence of pupils' work, teacher-pupil dialogue and a systematic approach to the diagnosis of strengths and weaknesses.

At Jasmine High School, Wigan, the recording of subject-specific achievement was the area of greatest development. Perhaps the most significant move in this area of the recording of achievement was the move made by a number of departments towards unit accreditation which appeared to gain momentum during 1986/87. These units of work were either selected from a bank of readily available validated units or prepared by staff at Jasmine. In the latter case, units had to be validated by the accrediting body, the Northern Partnership for Records of Achievement (NPRA), before being taught in the school. It was reported in June 1987 that mathematics, PE, French, home economics, music, humanities, science and English were involved in the scheme to a greater or lesser extent.

It would not be appropriate for us to advocate the universal adoption of one of these types, because to do so would not only be to judge one kind of RoA system as superior to another but to say that one form of curricular organisation is to be preferred. We hesitate to do this because we recognise that this would have far-reaching implications which schools may not be in a position to take on board and which may be more than is required for the development of a satisfactory RoA system. On the whole we have chosen to concern ourselves with what is possible and sufficient, although we do not wish to deny the possibility of an ideal for which schools might aim.

None of these types, for example, necessarily precludes the adoption of a wide definition of achievement which goes beyond the knowledge of subject content to include conceptual, practical, process, personal, social and attitudinal achievement. Neither does any type necessarily preclude a central role for formative processes of self-assessment, review and target-setting, although these are more clearly formalised in the last. Realistically, any decision about which of these types a school can be expected to adopt will need to recognise current organisation and starting points including the relative emphasis given to the role of form tutors, the need to communicate a synopsis of pupil achievement to outside audiences, and the development of formative processes within subject areas. Schools' ability to adopt one or another of these types will also be constrained by the availability of resources to enable fundamental change in their forms of organisation.

We now turn our attention to the different **conceptions of subject-specific achievement** which have underpinned recording systems in our case study schools. In this respect, our evidence suggests that systems developed to record subject achievement, within any of the types outlined above, pay attention to one or more of the familiar trilogy of knowledge, skills and attitudes.

By **knowledge** we mean propositional knowledge specific to the subject which is assessed by requiring pupils to demonstrate recall and understanding. Traditionally this has had a prominent role in education and, according to our evidence, is still an important focus for assessment, recording and reporting.

Attitudes constitute another aspect which has always been a prominent focus for recording in relation to subject achievement, in particular the interest in, and disposition towards, work within the subject as demonstrated by effort, commitment and motivation. Assessment of a pupil's 'effort' has had a long history in education in England and Wales although the assessment of attitudes **deliberately fostered within the curriculum** (such as those promoted within, for example, health education) is a different matter.

Skills constitute the third aspect and it is this area which has received particular attention in some schools and schemes, in keeping with the current prominence of skills-based approaches to curriculum and course design and development. Indeed 'skills' has become something of an educational buzzword, the meaning of which has been obscured by possible overuse. This observation applies equally to the term 'processes' to which it is closely allied, sometimes to the extent of being used synonymously. For the purposes of clarification, variations in the meanings attached to term 'skills' can be conceptualised according to definitions on three dimensions. On the first dimension, narrow definitions of achievement in terms of skills often focus on performance of some generalised operation (or process) devoid of reference to specific content and context (e.g. problem-solving) although more comprehensive definitions acknowledge that skill also involves knowledge of ends and the means to achieve them (propositional knowledge) and a disposition to use means to achieve ends (attitudes). On the second dimension, the concept of a skill developed in subject contexts can be restricted to what might be termed practical or craft skills (e.g. ball control in PE; manipulation in CDT) or it can be broadened to take in intellectual skills (e.g. critical analysis, investigation) and personal and social skills (e.g. empathy, collaboration). On the third dimension, skills can be singular and discrete (e.g. asking questions, hypothesising, collecting data, categorising) or they can be composite (e.g. investigating, researching) indicating different orders of skills in which higher order skills subsume lower order skills.

We do not wish to get into an extended discussion about the nature of skills, although we think it is worth noting that the place of skills in the curriculum is an area of considerable debate. Neither do we wish to advance one interpretation to the exclusion of others, although, as we shall argue, our evidence suggests that, at least for summary purposes, a broad definition on all of these dimensions is to be preferred i.e. that the skills of most interest are higher order, inclusive skills defined as practical knowledge in personal, social and intellectual areas as well as craft skills.

Although most of the schools which we studied implemented RoA systems which incorporated subject-specific knowledge, attitudes and skills, to a greater or lesser degree, a few chose to make skills (or processes) a principal focus for recording. In these contexts what was identified, assessed and recorded as skill or process achievement varied widely both between and within institutions, along the lines described above. A number of issues of special interest emerged but the greatest problems were experienced when skills were recorded as atomistic, general competencies devoid of reference to the content and context of demonstration.

At Cornflower Special School, OCEA, assessment frameworks developed at scheme level for both science and maths were used but with adaptations. These frameworks were built on process models and assessment criteria were framed in terms of pupils' skills in relation to these processes. Thus the mathematics assessment model was based on a concept of problem-solving and included the processes: starting, measuring, calculating, representing, checking, conjecturing, generalising, justifying and communicating. Similarly, the science assessment framework was 'based on a view of scientific activity as an empirical approach to problem-solving' involving the processes of planning, performing, interpreting and communicating. Each process was further sub-divided and process skills demanded of pupils were classified in an order of difficulty. Initially, all skills were classified according to four levels but this formal structure was abandoned in order to allow greater flexibility and to avoid artificially forcing criteria into four steps. In attempting to implement a recording process based on the OCEA assessment frameworks, teachers at Cornflower School encountered a number of difficulties arising from the special context in which they worked. First, pupils could not understand the assessment criteria which were used for recording. Secondly, for those with the most severe learning difficulties even criteria at the lowest level were pitched too high. For these reasons teachers found it necessary to help pupils translate criteria into language which they could understand, or make judicious selections of criteria from the OCEA models in order to avoid the negative effects of consistent non-achievement. Thus, for instance, the maths teacher began recording with just four simplified criteria: 'I can explain what the problem is about'; 'I can choose what I need'; 'I can follow instructions correctly'; 'I can use everyday units of measurement'. Although the assessment of process skills was the central element in OCEA G component recording, subject content covered was also formally recorded in maths. The G maths summary statements prepared in June 1987 included reference to subject content, whilst G science statements were purely concerned with the process skills which pupils had demonstrated i.e. without reference to content or context.

Validity issues were highlighted in relation to the assessment models used as the basis for G component recording, particularly in science. Without doubt they provided detailed frameworks which encouraged a very much more systematic approach to recording than had hitherto been the case.

However, the emphasis on processes led to recording in terms of skills with little or no reference to the context of learning. This meant that whilst, of necessity, skills were fostered in relation to content, pupils' knowledge and understanding of, say, magnetism or the properties of light were not included in recording at any summary stage. Thus, what was recorded reflected only one aspect, albeit an important one, of the curriculum as taught and learned. The OCEA science teachers' guide acknowledged this limitation but justified its emphasis on doing rather than knowing as a necessary counter to previous practice.

However, it remains the case that statements of process skills demonstrated are not easily understood without some indication of the contexts in which they were achieved; in particular they give little indication of degree of sophistication. For example, 'The student can make relevant observations of a general nature' was intended to describe a low-level skill but performance might be expected to vary according to what was being observed. Whilst moderation might standardise teachers' assessments on a criterion, this is no assurance that non-professional users of records would understand what was meant.

On the basis of this evidence, our judgement is that subject achievement which is recorded in terms of skills will need to include reference to the content and context of performance if it is to have validity and meaning. Whilst skills may have similarities in different contexts they are also distinguished by their content. It cannot be assumed therefore that they are entirely transferable.

Similarly, although it may be helpful to record the performance of numerous discrete skills (e.g. in comment banks) at the formative stage, the indications are that these need to be summarised in more holistic ways when there is a summative intention to report to other audiences. Thus ways need to be found to order achievements in some kind of hierarchy such that atomistic skills, if it is felt that there is value in recording them, are subsumed within more holistic categories at the stage of reporting. This is important because the constraints on space in any summary document create a need for selection, but limited selections of atomistic skills tend to trivialise and misrepresent achievement.

Most pupils interviewed by the evaluator at Sunflower High School, Lancashire, said that they liked the new (comment bank) end of year reports and the amount of detail provided, but had some criticisms. Two pupils interviewed remarked that some statements, e.g. those included in the mathematics RoA, were difficult to understand. In contrast, one pupil felt that some statements were 'babyish', e.g. 'X is able to throw the ball standing still'. Two pupils remarked on inaccurate assessments. For one pupil the statement 'X needs help while using the microwave' was included in the interim RoA but could not recall having used the microwave oven at all! Whilst most pupils thought that the interim statement should be entirely positive, some thought that areas needing improvement should have featured more prominently. Most parents who returned the questionnaire sent out as part of the evaluation remarked that they liked the interim statement because it was easy to read, well-organised and provided a fuller picture of pupils' academic and personal achievements. However there were criticisms that the information did not indicate how well their child was doing in relation to others in the year, that it was impersonal, and that some of the comments were 'silly' (e.g. 'X can catch a ball with two hands.').

Our evidence also points to a growing recognition that a wide range of skills is brought into play in learning within subject contexts. This range is likely to include intellectual, social and personal skills as well as manipulative skills. Indeed, the experience of teachers in some schools suggests that curricular opportunities for demonstrating social and personal skills are best provided within the formal academic curriculum where they arise naturally and are often an essential element (e.g. skills of collaboration).

There was some evidence at Rose School that, towards the end of the pilot phase, subject departments were beginning to develop recording frameworks which included social and personal skills, such as co-operation, as important aspects of subject-specific achievement. (The humanities department was especially notable in this respect.) The overall effect of these developments was to make any distinction between pastoral and academic recording increasingly untenable. Subject teachers and form tutors were both attempting much the same task and implicitly acknowledging, on the one hand, that social, personal and general learning skills are important for subject-specific achievement, and, on the other hand, that social and personal qualities cannot be considered devoid of context — and one important context is provided by the formal subject curriculum. The co-ordinator pointed to this as evidence that her particular aim to erode the pastoral/academic divide was moving to fulfilment. However, it is unlikely that such distinctions will totally disappear as long as a structural divide is perpetuated i.e. that schools continue to be organised in terms of pastoral curricula provided by form tutors, and academic curricula provided by subject departments. In the light of proposals for a national curriculum it will be important that social and personal aspects of subject achievement are taken into consideration in designing programmes of work and devising appropriate assessments if the insights provided by experience in schools like Rose are not to be lost.

On the basis of our evidence, our judgement is that a broad definition of skills appropriate for development and recording in subject areas should be encouraged, and that this should include skills of a personal and social kind which have an important bearing on subject-specific achievement. Indeed we feel that this is an important and valid way of approaching the recording of cross-curricular achievement and personal qualities, which is more specifically discussed in the next two sections.

Although many of our case study schools have expressed an interest in incorporating recording of cross-curricular achievement as a distinct element in their RoA system, and some have established working groups with this in mind, there has been comparatively little progress towards implementation in this area. This appears surprising but can be accounted for by the theoretical difficulties in defining cross-curricular achievement and by patterns of departmental organization which tend to inhibit cross-curricular deliberation in secondary schools.

(c) Cross-curricular achievement

In our case study work we encountered two basic approaches to understanding the term 'cross-curricular' and we feel that these need to be taken into account in the continuing debate on recording pupil achievement. First, there was the approach to cross-curricular which centred on the **pupil.** This approach was grounded on the assumption that general processes, concepts and skills can be identified which do not appear exclusively related to the substantive content of subjects. They are recognised to have a level of generality although they do not require some common agreed attempt across subject areas to foster the same skills in different contexts. Thus cognitive skills of investigation, social, personal and psycho-motor skills, and attitudes to learning, have all been referred to at one time or another as cross-curricular achievements. Such achievements may be developed, demonstrated and recorded in subject-specific contexts without reference to other subject areas. In some cases, as mentioned in the previous section, they may be considered **essential** to achievement in those contexts, in which case aspects of subject and cross-curricular achievement become indistinguishable and the term 'cross-curricular' becomes simply a way of indicating a level of generality beyond the particular. The substitution of terms such as 'general competencies' for 'cross-curricular skills' is significant in this respect.

At Hydrangea School, ILEA, the official definitions of a pupil's achievement which provided the basis for the project and to which staff often referred were those set out

11

in the Hargreaves Report, *Improving Secondary Schools*[1]. The report posited four 'aspects of achievement' which schools should aim to foster, to provide opportunities for pupils to develop and demonstrate, and to record in positive terms. The extended definitions of aspects of achievement III and IV overlapped considerably with the kinds of 'personal qualities' and 'cross-curricular skills' which have been explicit features of RoA systems in other institutions. *Improving Secondary Schools* explains aspects III and IV as follows:

achievement aspect III

This aspect is concerned with personal and social skills; the capacity to communicate with others in face-to-face relationships; the ability to co-operate with others in the interests of the group as well as of the individual; initiative; self-reliance and the ability to work alone without close supervision; and the skills of leadership.

achievement aspect IV

This aspect of achievement involves motivation and commitment; the willingness to accept failure without destructive consequences; the readiness to persevere; the self-confidence to learn in spite of the difficulty of the task.

In fact, in addition to being referred to as 'aspects III and IV', the kinds of skills, abilities and qualities which are encompassed by these definitions were also referred to within the school as 'personal qualities' and 'cross-curricular skills'. The term used to refer to these skills, abilities and qualities varied according to context. The term 'personal qualities' was used in the context of form tutors' writing of summative School Statements for the London Record of Achievement. The term 'cross-curricular skills' (and sometimes 'affective characteristics') was used in the context of a proposed 'cross-curricular profile of work related skills' (see 1.1.4 below). In the context of individual departments' formative subject Unit Credits and summative subject Credits, the terms 'aspects III and IV' and 'cross-curricular skills' were used virtually interchangeably. This variability in terms of reference, but stability in the skills, abilities and qualities to which the terms were understood to refer, would suggest that the sharp distinctions which have been made in other institutions, particularly between 'personal qualities' and 'cross-curricular skills', may be more a distinction between the kinds of contexts in which these characteristics may be developed, demonstrated and recorded than one between distinct types of characteristics.

The second approach appeared to centre on a notion of cross-curricular which related to **school organization**, and suggested skills which could be fostered, assessed and recorded in two or more departments and which were therefore identified as a shared objective.

At Begonia High School, EMRAP, a list of cross-curricular skills formed part of the first subject-specific RoA documents issued for third year pupils in February 1986. This list was obtained from another school and modified. The skills assessed by academic staff were: quality of work, approach to classwork, homework, understanding, oral work, presentation of work, conduct, progress, attainment in year, and attaintment in teaching group. Comment banks were provided for each area

assessed and academic staff were required to tick the most appropriate comment. Staff interviewed were critical of this section because it was felt that the skills identified were inappropriate to different subjects, that more than one comment could apply to the same pupil although the categories were supposed to be exclusive, and that the gradations between levels were difficult to refine. This section was removed following the first issue of an interim RoA statement for third year pupils in 1985/86.

In practice, there was sometimes movement from one approach to the other and the skills framing both kinds of approaches were often roughly similar.

---

[1] Inner London Education Authority: Committee on the Curriculum and Organisation of Secondary Schools *Improving Secondary Schools* ILEA (1984).

The strength of the first approach was that it allowed for the possibility that no two departments could foster precisely the same skill because skills have a knowledge content which is to some degree context specific, even though they may have much in common at a general level of analysis e.g. drawing conclusions from evidence. On the other hand, the second approach had the advantage of encouraging fruitful discussion and collaboration between departments. Ideally, some combination of these two approaches should be possible — they are not mutually exclusive. However, in the light of the evidence from our case studies that cross-department agreement is difficult to achieve and that transferability is a questionable assumption, our judgement is that the first approach has the greater potential in this area. There is a strong implication in this that the term 'cross-curricular achievement' is a misnomer for which an alternative should be perhaps be sought; it is perhaps significant that the DES policy statement refers to general learning skills.

(d) Personal achievement

From what has been said already readers will be aware that, in addition to extra-curricular and out-of-school interests, RoA systems in case study schools often included reference to personal and social achievements within subject-specific and cross-curricular recording. Indeed, as our analysis progressed, what had at first seemed to be clear conceptual categories became increasingly blurred. However, in keeping with their interest in portraying the whole person, all schools consciously developed a 'personal' element as part of their recording procedures, either as a separate record or within the curricular elements of the RoA. Although one can argue that all recording is personal because it focuses on individuals, schools in pilot schemes appeared to use the term in a more restricted sense to refer to achievements, experiences and qualities which were assumed to be pupil-specific rather than curriculum-related. Sometimes the reference was to personal qualities or attributes and sometimes it was to personal skills. The distinction is an important theoretical one but need not concern us here because in no case was there any interest in recording immutable individual characteristics which could not be fostered, developed or influenced by educational experiences. It is interesting that what in one place was labelled a personal quality (initiative, reliability, punctuality) was elsewhere described as a skill. On the whole any lists of such qualities or skills developed by schools or schemes to guide recording exhibited a remarkable degree of similarity. The proscription of comments on a pupil's honesty was widely advised; and the need to include some comment on attendance was also usual, although this was neither a quality or a skill.

At Daffodil Comprehensive School, Wales, 17 categories were itemised on one side of the single A4 card for the Record of Personal and Social Development: attendance, reasons for absence, punctuality, relationships with other pupils, relationships with adults, appearance, school activities, courtesy, consideration, co-operation, reliability, responsibility, self-discipline, assertion and resilience, oral communication, presentation of work, health as observed. A small space was provided for 'other relevant information'. Each category had two, three or four statements alongside it, e.g. 'excellent', 'room for improvement', 'cause for concern' and so on. These statements were envisaged as mutually exclusive, if not strictly hierarchical, and it was only possible to select one of the options at any one review (although 'half-way' compromises were sometimes reached between some pupils and tutors).

A stated aim of PSE diaries introduced as part of the RoA scheme at Geranium School, EMRAP, was to train pupils to talk about themselves in terms of their personal skills, personal qualities, strengths, weaknesses, etc., since older pupils at the school were regarded as being hesitant, embarrassed and untrained in doing this. The idea was that, within PSE lessons, pupils would develop skills and qualities such as dealing with people, speaking, sensitivity, sociability, working with those in authority, leadership, responsibility, ability to work in a team, self-awareness, self-confidence, flexibility, initiative, resourcefulness, attitude, enthusiasm, reliability and perseverance. Pupils would recognise the fact that they had demonstated such positive qualities in the particular PSE lesson and write about these in their diaries.

In the early months of the project at Marigold High School, Wigan, a list of qualities was drawn up between the RoA steering committee at Marigold and the head and assistant head of first year, along with a bank of statements for each quality to be assessed. The agreed list of qualities was: ability to cope with transition to high school, attitude to work, reliability, uniform, attendance and punctuality. However, following discussion with a member of the Wigan RoA team, there was, in the eyes of Marigold's RoA steering committee, a very welcome shift away from the idea of teacher assessment of personal qualities to a system whereby pupils would provide evidence of various interests and activities from which the reader could infer personal qualities possessed by the pupils. As it was felt that this now revised section of the RoA was covering areas of pupil interest as well as achievements it was renamed Personal Interests and Achievements.

Our evidence suggests that schools regard personal qualities and/or skills as an important aspect of achievement which deserves recognition within the RoA because personal and social development is held to be an important purpose of education. However it does not seem to us that universal agreement over what counts as personal achievement should be considered a pre-condition for objectivity, any more than it is in any other area (language, mathematics or science etc.). Criteria are, of course, implicit in the public form of discourse that teachers participate in for assessing such development. In other words the kinds of categories and criteria which have emerged through trial and review in the DES pilot phase are likely to have value in a record of achievement. As long as they are publicly accessible they are also likely to be used as starting points for discussion in other schools wishing to develop recording systems appropriate for their own special contexts. If criteria emerge and evolve in this way it is unlikely to be necessary to prescribe a list of qualities or skills to be recorded.

However, experience in pilot schools strongly suggests that, in order to be fair, personal qualities or skills should be recorded with sufficient detail of the content and context of demonstration to indicate that what is recorded is bounded achievement rather than fixed and immutable personality traits. There was also a considerable degree of agreement within pilot schools that comments should only be made on those personal qualities and skills which can be demonstrated in relation to the school's educational programme. This of course has implications for curriculum planning and demands that opportunities for fostering personal qualities and skills should be part of curricular and/or extra-curricular provision. Some of the pilot schools we studied made deliberate efforts to enhance curricular provision in this respect.

At Magnolia Girls' Grammar School staff felt that there was little validity in commenting on specific personal qualities if girls had not been provided with adequate in-school opportunities to demonstrate them e.g. leadership. For this reason a deliberate effort was made to create areas for responsibility and/or service. Thus a list of possibilities was created (e.g. sorting archive materials for the history department) and girls where given an opportunity to volunteer.

In our discussion of aspects of achievement we adopted a categorisation which often framed professional discourse in case study schools. However, the terms 'extra-curricular', 'subject-specific', 'cross-curricular' and 'personal' have strong associations with the way the curriculum is **organized** in secondary schools and this may be misleading. At various points we have said that these categories are not distinct or mutually exclusive in terms of what constitutes achievement. For example, personal qualities are not distinct from social skills since they may only be demonstrable in social contexts; what are called cross-curricular skills, such as co-operation and taking responsibility, are often indistinguishable from formulations of personal achievement; and both so-called cross-curricular and personal achievements are integral, even essential, to subject-specific achievement.

We are impressed therefore by the way in which the Hargreaves Report, which influenced the development of the London Record of Achievement, avoided tying the discussion of aspects of achievement to the organisational structures of schools. Instead it created categories of analysis labelled simply aspects of achievement I, II, III and IV. Aspect I related to propositional knowledge as assessed by examinations with an emphasis on an individual's ability to memorise and organise content; aspect II related to practical processes and skills involved in applying knowledge e.g. problem-solving, investigative skills or manual skills; aspect III related to personal and social skills such as communication, co-operation, initiative, self-reliance and leadership; and aspect IV related to positive attitudes associated with motivation and commitment.

According to this analysis it is possible to see that what we have referred to as subject-specific, cross-curricular and personal achievement can incorporate more than one of aspects I to IV. Indeed the authors of the Hargreaves Report argued that subject-specific achievement properly encompasses all four. The fostering of all these aspects of achievement therefore transcends organizational boundaries; recording in relation to most aspects can, and perhaps should, take place in relation to all elements of the educational programme of schools. Although the practice may fall short of the theory, as indeed our evidence suggests, this was the thinking underlying the development of units and unit credits in ILEA. Insofar as units and unit credits can in principle be developed to cover all aspects of achievement in relation to any curriculum unit – academic or pastoral – we feel the approach has much to recommended it. Of course, if Hargreaves' four aspects were to be used as straitjackets for the development of skills in every curriculum unit, classified under four reporting headings, rather than simply as categories for analysis, then their effects could be deleterious. The advantage of the Hargreaves approach, as we see it, is that it provides a reminder of the comprehensiveness and multi-dimensionality of subject achievement, and has some potential to erode the hard-and-fast academic and pastoral divide which has been erected and sustained by essentially organisational boundaries. The implications for curriculum planning, for teaching and learning, for assessment and recording and for in-service and other forms of support are of course substantial. However, our evidence leads us to the conclusion that this broadening of the scope of recording across the whole curriculum is the only valid way to portray the educational experience of the whole child within the existing organisational structures of schools.

Having examined what aspects of achievement were covered by records of achievement systems in our case study schools, we now turn to a detailed examination of the **processes** of recording which have been developed and used in schools. The term 'processes' has frequently been used in pilot schemes and schools to mean those activities, carried out by pupils and/or teachers, which have been devised in order to realise the central principles underpinning the RoA pilot programme. Indeed for most people working in this field the processes of recording have collectively been the hallmark of the RoA movement, since it is in the domain of these recording processes that the innovatory quality of RoAs has been most often identified. It is also in this area that most claims have been made for the capacity of RoAs to stimulate fundamental change.

For these reasons we have given processes of recording particular attention in our case study work. We have tried as far as possible in this report to discuss the processes of recording separately — not only from the kinds of achievement covered by those processes (which we examined in the previous 1.1), but also from the documents which have often been the outcome of these processes (which we look at in the following 1.3). In reality however the three areas are closely inter-related.

In the first four sub-sections we discuss the four main processes of recording which have characterised RoA systems in our case study schools. We first look at those processes which have enabled pupils to account for their own achievement, including continuous recording, statement-writing and self-assessment (1.2.1 **Pupil self-accounts**). We then turn to those processes developed as part of RoA systems in which teachers have accounted for their pupils' achievement, and in which teachers have worked essentially separately from pupils, such as the preparation of teacher-written statements and the application of new frameworks for the assessment of pupil achievement (1.2.2 **Teacher accounts**). Next we consider the wide range of teacher-pupil discussions which have been conducted to facilitate pupil self-accounting processes, to review progress and achievement and to assist with the production of various RoA documents (1.2.3 **Teacher-pupil discussion**). Fourthly, we examine the ways in which teachers have taken responsibility for collecting, collating and aggregating RoA-related assessments and accounts emanating from the first three processes (1.2.4 **Teacher co-ordination of records**). We then include a final overview section (1.2.5) in which we draw together our evidence relating to themes which cut across the various processes. In that overview we consider in particular the target-setting and diagnostic functions of recording processes, and the effects of recording processes on pupils' motivation and personal development.

1.2.1.
Pupil self-accounts

In this section we discuss three ways in which we have observed pupils portraying their own achievements. To describe these three procedures we have used the terms 'continuous recording', 'statement-writing' and 'self-assessment'.

(a) Continuous recording

We use the term 'continuous recording' to cover those forms of recording in which pupils described personal, and sometimes academic , achievement and experiences on some kind of regular diary-keeping basis. The continuous recording by pupils of their achievements and/or experiences, whether in or out of school, was a feature of many RoA systems observed. Continuous recording was usually but not always conducted in free prose, and pupils had a measure of choice over the content, although topics were sometimes supplied by teachers. Continuous recording processes did not refer to specific dimensions of learning or to levels of achievement, and was not addressed to a public audience, apart perhaps from the pupil's teacher or form tutor. Much of this type of recording took place from the tutorial vantage point, but this does not mean that the academic side of school life was necessarily excluded. When writing personal accounts pupils sometimes referred to their achievements and experiences within the academic curriculum and were often encouraged to do so (see 1.1.2).

Continuous recording was known by a range of different names in case study schools, for examples:

• personal logbooks or files (Nasturtium Tertiary College)

• logbooks (Buttercup RC School and Sunflower High School)

- diaries (Rose School, Geranium High School and Wisteria School)

- box files with P Component booklets and tapes (Cornflower Special School)

- 'profile' file (Campion Upper School and Hawthorn RC School)

- 'This is Me' diaries (Bluebell School)

- 'Think Books' (Columbine Middle School)

- 'Interests and activities outside the classroom' pro-forma in pupils' personal folders (Daisy High School)

In many cases pupils were expected to structure their entries in these records around themes, topics or questions provided by teachers or RoA project teams. The most common focus for continuous recording was upon extra-curricular and out-of-school activities, interests and experiences. Schools placed great value upon such recording early in the pilot period, since it held the potential for:

(a) providing a basis of evidence which was compiled as relevant events transpired and upon which subsequent summaries of achievement (e.g. statements) could be composed, thus avoiding the necessity of relying on memory; and

(b) enabling teacher-pupil discussions which focused upon a more rounded portrayal of the pupil, thus providing teachers with a broader understanding of their pupils as people and strengthening teacher-pupil relationships.

Although the principles which underpinned these forms of continuous recording are still alive in schools, we must report that, for a number of inter-related reasons, all of our case study schools have virtually abandoned the process of continuous recording as it was originally formulated. It is true that some pupils took to the process of regular recording of personal interests and activities with great relish. Interestingly, these pupils tended typically to be girls (see 1.5 below). A number of these pupils were already accustomed to keeping diaries of their own. For the majority, however, interest waned rather quickly. Pupils reported their difficulty in 'thinking of anything to write', and that despite assurances to the contrary from their teachers they began to experience diary-writing as yet another form of 'work' they 'had to do'. The medium of writing, the unchanging focus, and the potentially revealing nature of entries were all cited as reasons for their reluctance. These factors were compounded by the difficulties experienced by teachers (usually form tutors) in trying to establish a regular routine of making entries — either in the classroom, when there were so many other demands on limited classroom time, or at home.

These factors were closely associated with the general trend we have observed in case study schools, and which we discuss elsewhere in this report (e.g. 1.3.2), away from 'purely' formative procedures, and towards procedures which incorporate some kind of public reporting function. It is difficult to say whether this trend has been the **result** of the gradual discontinuation of regular recording identified above, or part of its **cause.** In either case the early attempts by case study schools to introduce continuous recording gradually gave way, either to the production of intermittent statements (see next section), which served an explicit summarising and reporting function, or to a more varied programme of recording in which the format, purpose and focus changed frequently.

Many pupils at Periwinkle School, OCEA, mentioned the problem of completing their personal files. In some cases they felt that the time they were given for recording was not necessarily the time when they felt moved to write something down and the problem was one of having access to the file at the appropriate time. In other cases it is the lack of something to say which is the problem.

Pupil: Me, well it depends what you're writing about. I enjoyed writing about the pets — you have a whole double lesson and you've got to write something down or else! If you can't think of anything you've just got to write anything down. It's quite noisy in the classroom, they've been talking amongst their friends because they haven't got anything to write.

Researcher: So if you've got something that's happened to you that you think is good, then you like writing it down?

Pupil: Yes, like if you go on holiday you feel you really want to write it down but other times it's just so boring. I haven't had the chance to write about the pantomime, I still haven't had the chance to write it down because we haven't had a tutorial for so long. I've still got to write this down, but some people they don't have anything to write down and they still haven't finished doing it. You get a folder to write something new in and then you're flipping through and you think, 'Oh I haven't finished that off, I can add more onto that,' things like that. It just reminds you if you do something else like that it reminds you to write about things.

Older pupils, in one group at least, had no set hours for personal recording but were expected to write things down when they had the chance. Some felt that the recording should have more of a focus such as views on a particular topic in Current Affairs. Some felt that the file was just given at the beginning of the year and not explained, they felt belittled by the whole tutorial programme and regarded OCEA in the same light: as something they did when they had nothing better to do. There seemed to be a general problem in writing regularly about personal experience and reflections and it would appear that this cannot be sustained throughout a whole school career. This problem was undoubtedly compounded by pupils' lack of understanding of the purpose of the exercise. Some older pupils felt the entries were not specific enough to be useful at an interview and were not even aware of the forthcoming summative document. Younger pupils exhibited similar misunderstandings. (N.B. 'Well, if we haven't got anything important that he had planned — then we just do our folders.')

For most pupils in the class studied at Columbine Middle School, entries in 'Think Books' gradually became less and less frequent, and less and less reflective, as time went on. This was partly due to shortages of class time for writing, and the consequent difficulties and variations associated with sustaining an activity of this kind at home. In the second year of the pilot, most pupils had virtually stopped making entries in their Think Books altogether. This was partly due to the fact that they experienced a change of form tutor, who gave less prominence to Think Books relative to other kinds of tutorial work such as group discussions, role play and worksheets on tutorial topics, and partly to an element of staleness and boredom which

had set in. At the end of the second year this tutor did acknowledge, however, that the process of producing an end-of-year statement had been more onerous than it might have been, as a result of too little time having been given to recording through the year. He regretted the fact that the process of producing the statement had not arisen directly from the process of regular recording. Another effect was that some pupils used their statements more formatively than they might have done had they had more frequent opportunities to record their experiences through the year. One pupil, for example, used his draft statement to describe an upsetting event from the previous weekend, rather than to reflect back over the year.

The only form of continuous recording which, according to our evidence, has been sustained in essentially its original form, has been the more administrative type in which for example extra-curricular and out-of-school activities are simply added to some kind of pro-forma at regular intervals.

At Daisy High School, Wales, pupils were encouraged to record regularly their

interests, experiences and achievements outside school on an 'Interests and

Activities' pro-forma in their personal folders. Such activities may have included hobbies and interests, membership of organisations, clubs and societies, charity work, work experience, helping at home and so on. This form of recording centred on factual information, briefly entered on the pro-forma, rather than, for example, on subjective accounts of feelings, or descriptions of events in extended prose, though these may have been discussed in tutor-pupil interviews.

Whilst it would be inappropriate for us to come to a judgement as to which of these directions is most advisable, we would emphasise the almost universal movement away from the process of continuous recording, and inform those schools wishing to establish some kind of recording process that the only kinds we have observed to have been sustainable over time have been those in which the format, purpose and focus changed frequently, or those which were essentially administrative.

One consequence of this movement away from the original concept of continuous recording is particularly worth mentioning here. It is that the issue of confidentiality, which preoccupied projects at the early stages of the pilot, has become considerably less problematic, since it has become increasingly common to combine personal recording and public reporting functions. Although naturally there has been an accompanying loss of intimacy in some pupils' entries, this trade-off is one which most schools seem prepared to accommodate in the interests of rationalising time and effort, and of capitalising on pupils' greater support for more audience-directed forms of recording.

(b) Statement-writing

By 'statement-writing' we mean the production by pupils of free prose accounts which summarised achievements over a specified period of time, usually a single school year or the whole of a pupil's school career. Like continuous recording, statement-writing has usually been carried out from the tutorial vantage point, though in many cases pupils have been encouraged to reflect on their achievements within the formal academic curriculum as well as upon their activities, interests and experiences outside the classroom. A large number of the one-to-one discussions we have observed between teachers (usually tutors) and pupils have been directed towards the production of such statements (see 1.2.3).

We noted above the trend we have observed generally away from 'purely' formative processes and towards procedures which combine a formative recording function and a summarising public reporting function. One result of this trend (or possibly part of its cause) has been that earlier processes of continuous recording have gradually given way to a growing tendency to produce interim statements of achievement, usually at the end of the school year and usually supported by pupils' form tutors. Schools have increasingly come to rely on statement-writing processes to achieve many of the formative functions which it was originally hoped continuous recording processes would serve, such as the creation of an evidential base (though on this revised pattern this base is now directed towards summative RoAs) and the strengthening of teacher-pupil relationships.

For the most part pupils at Cornflower Special School were withdrawn to the 'office' for one-to-one discussions on the P component although the OCEA assistant who conducted them had to be opportunistic about choice of venue. Usually, the OCEA assistant used the observations she had noted in lessons as the basis of discussion, together with pupils' self accounts in their 'P booklets' and, on one occasion, a pro-forma of specific questions which pupils had been asked. The overt purpose of these review sessions was to prepare interim summary statements but since these statements were purely descriptive summaries for the pupils alone their function was mainly formative. They therefore gave pupils a legitimate opportunity to open up about themselves, their families, friends and their school experiences. Whilst they may have been a cathartic experience for some, there was no evidence that these discussions were used for explicit diagnostic and counselling purposes. The OCEA assistant thought pupils saw her as the sort of person to whom they would turn for 'motherly' advice but she was not a teacher, nurse or social worker so avoided assuming a professional role. Whilst some opportunities were undoubtedly lost others were perhaps gained since pupils did not have to worry about which hat the assistant was wearing at the time. Indeed their statements indicated that they felt relatively free to say what they wanted to — even to grumble about the headteacher.

At Sunflower High School, Lancashire, City and Guilds-style logbooks were rejected after a year and continuous recording ceased to be a feature of the pastoral scheme. End of year statements were produced by pupils throughout the pilot project. In one such statement the pupil wrote about what he had enjoyed in school, what he felt he had learnt, and details of his extra-curricular and out-of-school activities. At the end of the 1986/87 year these end-of-year statements were issued for first and second year pupils as part of an interim RoA which also contained subject-specific RoAs developed during the project.

This gradual fusing of the initially distinct processes of continuous recording and statement writing is simply one example of the more general trend, which deserves to be acknowledged formally at several points in this report, towards a blurring of the distinction between formative processes and summarising processes. This trend seems to us to constitute a significant outcome of the pilot programme, and one which should be recognised particularly by schools setting out to develop RoA processes.

One consequence of this growing dependence on statement-writing to serve formative functions is that the task of preparing statements at the end of the school year has in some instances been recognised as more onerous when that task is attempted without the benefit of regularly-kept records, since it demands that pupils rely on their memory for evidence of achievement, and since it requires pupils to prepare their statements without regular practice of reflecting upon their achievements (see the example from Columbine Middle School in (a) above).

Another consequence is that pupil-prepared summative statements, which were a feature of many summary RoAs in our case study schools, have gradually become much more explicit 'summaries of summaries', which may ultimately make the transition from formative recording to final summative reporting less problematic than was originally feared.

One theme which recurs repeatedly in our data relates to the difficulties many pupils have had in gaining a clear understanding of the nature of the statement-writing task. Whilst this has perhaps been more common of younger pupils, it has by no means been confined exclusively to them. In one sense this poor understanding of the process is hardly surprising, since it represents a significant departure from pupils' normal experience of school tasks. Pupils may therefore require a great deal more experience of this kind of activity before gaining a clear understanding. On the other hand, their imperfect appreciation of the **purpose** of the exercise, and of the **audiences** to which their statements might be addressed, seems to us to have prevented them from approaching the task of statement-writing with enthusiasm, and from producing statements which will serve the multiplicity of functions which teachers envisage for them.

Vague notions of 'employers', or 'college', have characterised many pupils' responses to our questions about 'who might want to read the statement'. For younger pupils possible end-users are probably too remote an audience to inspire motivation and conscientiousness. Some of the most highly motivated pupils to whom we have spoken have seen their statements as being addressed essentially **to themselves.** They have written their statements in order to read about themseves at this age when they are older. Many pupils seem to have an extraordinary capacity to plan for nostalgia, and capitalising on this is probably as valid a basis for establishing motivation and conveying the long-term purpose of the exercise as any other. We fully acknowledge the difficulties faced by colleagues in schools when attempting to inculcate in their pupils a clear sense of purpose and audience in relation to statement-writing, but must stress nevertheless the necessity of finding successful ways of doing so if this process is to be sustained.

(c) Pupil self-assessment

We use the term 'self-assessment' to refer to specific judgements or ratings made by pupils about their achievement, often in relation to teacher-designed categories. These tended to be more abbreviated than continuous records or statements, and were often expressed in grids rather than in free prose. We have observed self-assessment taking place in relation to personal qualities, cross-curricular skills and (perhaps most often) subject-specific achievement. These self-assessments, together with teachers' assessments, often provide a basis for teacher-pupil discussion.

For many of our case study schools, pupil self-assessment has constituted one of the most significant elements of the RoA system, holding as it does the potential for realising a number of central RoA principles. Moreover, unlike some of the other processes which have been developed as part of RoA systems, it promises eventually to reduce the time demands of RoAs on teachers whilst serving valid educational purposes, although in the short-term the reverse sometimes appears to be the case.

> Pupil self-assessment was one of the principal features of subject profiles at Bluebell School, Dorset. Indeed, the pupil self-assessment element of coursework profiles constituted such a prominent and innovatory development at Bluebell that most pupils interviewed by the evaluator understood the term 'profile' simply to mean the provision by subject teachers of an opportunity for pupils to assess themselves in one way or another. Pupil self-assessment was seen by many members of staff to hold the key to many of the project's objectives, including that of improving teacher-pupil relationships and increasing pupils' sense of ownership and personal responsibility for their own learning.

> Self-assessment was a central feature of the RoA system developed at Rose School, and the term was applied equally to the process of identifying evidence of personal achievement and qualities and to assessment of performance on subject-based criteria. In relation to the latter there was a remarkable degree of homogeneity in view of the fact that departments developed their own recording systems more or less independently. Departments that produced RoA-type schemes mostly included a substantial element of student self-assessment. This often included recording work covered, marking work on the basis of a teacher-generated marking scheme, or completing a self-assessment sheet for use as a formative record in discussion.
> The first assessment framework to be developed was in French, although this went through several experimental stages in the pilot phase. Essentially it involved dividing the course (for first and second years in the first instance) into two or three week topic units covering four skills (or profile components): speaking, listening comprehension, reading comprehension and writing. (The Tricolore course lent itself

particularly well to this use.) At the beginning of each unit pupils were given the objectives for the three week period and told what they would learn. At regular intervals during the unit of work — mostly every lesson — pupils were asked to assess themselves and/or test each other in relation to skills and the topic area. At first it was hoped that most assessment would be self- or peer-assessment but this proved unrealistic because self-assessment needed to be backed up by lengthy teacher comment. This made too great demands on teacher time and was simply not done. Eventually a decision was taken that pupils should only be expected to test each other in relation to oral work. Moreover, with the publication of national curriculum proposals for testing at 7, 11 and 14, the department felt that it would fail pupils if it did not train them for tests; therefore teacher-administered and marked tests were re-introduced although they were integrated into the normal pattern of lessons and were not used to compare one pupil with another but to encourage pupils to compare their own performance over time. Test scores were recorded by pupils on record sheets and short diagnostic comments by both teacher and pupil were added and discussed before being signed by both. These sheets were then transferred to formative RoA files where they were available for revision purposes and for writing interim summary and final reports. It was intended that they should become the property of pupils at the end of their education in the language.

This return to a rather more conventional pattern of assessment is interesting because initially the Head of the Modern Languages was highly committed to an approach based on almost total self-assessment and her own enthusiasm was clearly communicated to pupils in her own classes which appeared to make it work. However, in formulating a departmental policy she was obviously aware that she had to carry others with her who were perhaps less committed. Problems with pupil self-assessment were also experienced elsewhere. For example, the Head of CDT said that some self-assessments were difficult for some pupils to make and, lacking formal standardisation, it was feared that they would fail to gain external credibility. Furthermore, self-assessment took considerable time in lessons and required a re-ordering of priorities which some staff questioned.

We have, in our case study work, encountered a number of imaginative uses to which pupil self-assessment has been put. Virtually without exception, all of these uses fit into one of three categories: feedback, specific judgements and self-evaluation.

First, **feedback** is, strictly speaking, more of a consumer's review of particular aspects of a course than a self-assessment as such, except insofar as pupils are invited to reflect critically upon their personal response to the course and to think analytically about what aspects of the course engendered the response. Feedback also includes pupils' responses to their teachers' comments on profiles.

Questions on the PE/expressive arts profile used with second year pupils at Jasmine High School included:

Q7: Do you think the course has helped you to develop in any of the six areas above?

Q8: Can you make helpful suggestions for improvement?

Second, **specific judgements** are what is probably most commonly understood by the term 'pupil self-assessment' — that is, pupils' rating or estimation of their attainment or competence with respect to a specified aspect or level of achievement.

In all curricular areas at Nasturtium Tertiary College, Essex, the approach to recording was basically similar. Students were expected periodically to record evidence of achievement under criteria or headings identified by staff and then assess their own performance according to four levels of achievement: Level 1 — no measurable achievement; Level 2 — achievement with help; Level 3 — achievement on own; Level 4 — exceeds course requirements. These assessments were then expected to form the basis of discussion with subject lecturers and course tutors as appropriate.

A form of 'specific judgement' was used at Daisy High School in relation to the 'Personality and Attitude' comment bank (for the recording of personal and social achievement). Pupils were asked to assess themselves on the qualities described in the bank by selecting those comments which they felt applied to them. Comments selected by pupils were then compared with those selected by their tutor in a one-to-one discussion.

Third, **self-evaluation** refers to those instances when pupils are invited to appraise their performance in a mainly diagnostic way — that is, to reflect upon their strengths and weaknesses in relation to specific work and to give thought to ways in which they could in future improve their performance or derive more out of the work, but not to assess themselves in terms of specific aspects of achievement or levels of learning which were given priority by teachers.

Questions taken from a total of 24 questions appearing on an English report at Sunflower High School included:

What do you think you have learnt?
How could you take this topic futher?
Would you like the work to be assessed? If so, how?
How much time were you given to do this work?

Was it enough?
What did you find particularly satisfying?
What did you find particularly difficult?
Did you learn anything about yourself and/or others?
Did you understand what you were doing and why?

In the home economics department at Marigold High School pupils completed self-assessment sheets at the end of every lesson, assessing themselves on, for example, whether they chose the correct equipment for the task, whether they used equipment properly and whether they felt they did their best. As well as enabling pupils to evaluate their own work on a regular basis, the sheets were also seen as a method of providing pupils with a self-assessment record upon which to draw when completing the RoA.

Evidence we have collected highlights problems experienced by pupils in making their assessments.

We have evidence that pupils are strongly influenced when making assessments of themselves by their perceptions of the kinds of assessments teachers will find acceptable.

(i) Superficiality

The difficulty most often cited by staff at Bluebell School arising from their attempts to introduce and promote pupil self-assessment was that of encouraging pupils to reflect critically on their own learning. Staff frequently found that, on the one hand, pupils' comments tended to be superficial if they were not given detailed guidance on the coursework profile. Pupils' often responded to questions such as 'What did you learn by doing this assignment?' or to prompts such as 'Pupil Comment:' with only two or three words. Comments such as 'boring' and 'I need to try harder' quickly became the bane of many teachers' efforts to include self-assessment in their profiles. On the other hand, staff found that pupils tended to reply to more detailed prompts simply by re-phrasing the question in the form of an answer, thereby gearing their self-assessments towards meeting the perceived expectations of the teacher. Members of staff expressed the concern that teacher prompts which were too specific and closed-ended would shape pupils' comments and prevent them from real reflection.

Those members of staff who felt that they had begun to overcome this double-bind reported that pupils appeared able to reflect most acutely when (a) prompts and review questions were directly linked to specific aspects of their work rather than to skills which were too abstract for them to understand; (b) self-assessment was a regular and integrated feature of coursework rather than a periodic stock-take after the work had been done; and (c) the process of self-assessment and the terms used were a regular topic of discussion in the classroom rather than something confined to coursework profiles.

In the art department at Marigold High School, a double lesson was allocated to the pupil task of preparing the 'additional comment' section of the art RoA. The head of department reported that he was probed by pupils for information on what sort of things to include in this section. Pupils responded by answering the specific questions he had given as suggestions of what to include and he therefore questioned the validity of some of the statements produced. This early RoA scheme was replaced by one which was completed as work was in progress and which promoted a more personal response from the pupils. At the time of writing this scheme was in use and the problems of prescriptive responses was felt to be no longer applicable.

The head of textiles at Geranium School reported that she had tried out a self-assessment exercise with second year pupils, getting them to consider what they had enjoyed about the lesson, what they thought they could do well, etc. She found however that the pupils seemed to lack the skills of reflexivity, making comments such as 'I didn't like that.' She also reported that some pupils did not always take the exercise seriously and gave flippant responses. Consequently she found this exercise a pretty negative one but said that if the department could devise a pupil evaluation sheet that gave a positive response they would look to include this as part of an interim RoA statement.

(ii) Modesty   The difficulties described above may be compounded by the fact that pupils report that they underestimate their achievements in order not to lose face if they are subsequently downgraded. Teachers seem to attribute this latter phenomenon to the unwillingness of some pupils to appear boastful (cf. 1.2.3).

Few of the self-assessment processes introduced by departments at Bluebell School were based on quantitative ratings or grades. Indeed, the movement away from quantitative assessment processes and towards descriptive assessment was one of the central features of the RoA project. Although some departments continued to use grades, this emphasis upon descriptive assessment obviated to a considerable extent the problems of underestimation of achievement which have been associated with self-assessment processes in other institutions.

With one exception, all six fourth year girls at Magnolia Girls Grammar School whose RoA experience was studied in some depth, assessed themselves more harshly than did their teachers. The girls concerned admitted that they underestimated themselves and gave several reasons. First, although they completed the grids individually their friends could see what they had written which discouraged them from being boastful. Secondly, they said that they would be embarrassed if they found that staff had rated them lower than their own estimate and they consequently had to 'agree' to a lower rating. One girl said: 'You don't want to mark yourself too high in case they mark you lower.' Their form tutor claimed that an over-modest approach to self-assessment was a common phenomenon across the whole group to which these girls belonged. Contrasting this with her knowledge of German students she thought this might be a peculiarly British trait. Some of the girls thought it might also be a female tendency since they perceived boys to be much more boastful. The Headmistress also supported this with her own observation: 'In my experience girls (and indeed female teachers) consistently undersell themselves. In a self assessment/partner assessment exercise used (in Life Skills) involving only pairs of friends the vast majority over several years have marked themselves down on academic ability, appearance and figure'.

A further problem arises from the fact that pupils appear to have no clear frame of reference when making their assessments. Although presented with specific categories or criteria for assessment, many judge their achievement in relation to their perception of the range of achievement in their teaching groups.

(iii) Persistence of norm-referencing

One very articulate pupil at Magnolia Girls Grammar School described how she ranked herself, using her usual teaching group as the reference point, before making a judgement about the extent to which she had satisfied a criterion. She was not prepared to say that she had fully achieved the criterion if, in her estimation, there were others in her group who had performed better. She conceptualised the performance of her peers as above-average, average or below-average, and mentally placed herself in one or other of these sub-groups when considering each category on the self-assessment sheet. In other words she judged her achievement in relation to the distribution of achievement on the criterion (norm-referencing). rather than simply judging her own achievement on the criterion (criterion-referencing). She used this description of the process by which she arrived at a self-assessment to account for the discrepancy between her estimate of herself and the estimates of her made by staff. She thought that whereas she used her teaching group as her frame of reference (she was mostly in top sets), staff probably had in mind the whole year group when they tackled the task. In other words she assumed that they also operated some form of norm-referencing but across a slightly different population.

At Begonia School, one pupil reported that he and other members of his class had found the self-assessment exercise in geography difficult because some of the language was difficult to understand, e.g. 'can hypothesise'. They were aware that the skills increased in difficulty and that basic skills were on the left hand side of page and more difficult skills were on the right hand side. The teacher felt that when making self-assessments some pupils placed themselves along a continuum from poor to good rather than studying the statements closely to see which most accurately described them. In other words, because the language in the statements was inaccessible to them they norm-referenced rather than criterion-referenced their achievements.

The illustrations raise the question of whether pupils are able to assess themselves accurately and in a way which is free from understandable anxiety.

Pupils have a natural inclination, through years of experience within the educational system, to base their views of their own achievement on perceived teacher expectation, on perceptions of what is socially acceptable, and on conceptions of 'average' relative to a familiar group. What seems clear is that, if pupils are to become more autonomous and discerning in their assessments of themselves, teachers must provide abundant and supportive opportunities for them to gain practice and experience. In this respect the introduction of self-assessment processes probably requires an investment of a great deal **more** time by teachers, rather than representing a time-saving innovation, at least in the short term.

1.2.2

Teacher accounts

Many of the RoA systems we have studied have incorporated processes in which teachers have accounted for their pupils' achievement, and in which teachers have worked essentially separately from pupils. Such processes include the preparation of teacher-written statements, mainly but not exclusively at the final summative stage, and the application of frameworks for the assessment of pupil achievement. In an important sense two processes of teacher-accounting closely mirror the two central processes of pupil self-accounting described in the previous section, i.e. statement-writing and self-assessment. The two teacher-centred processes are usually more similar to each other than are the two pupil-centred processes, in that many of the new forms of teacher assessment are, in their attention to detailed evidence and description, more like 'statements'.

Perhaps it needs to be said that there is nothing innovatory about teachers writing reports on pupils or assessing pupils' achievement: these are tasks which have always occupied teachers. In this context it is worth asking what the essential differences are between the production by teachers of statements within RoA systems and the production by teachers of conventional school reports, and between the application of frameworks for the assessment of pupil achievement within RoA systems and more conventional assessment processes. One answer to this question, however partial, is that neither of the new processes are, at least in their ideal form, carried out exclusively by teachers: almost by definition they include elements of pupil self-accounting and teacher-pupil discussion.

In all cases the explicit purpose of the teacher-pupil review sessions at Hawthorn R.C. Upper School, Suffolk, was to discuss the content of the final RPA Statement. All pupils were asked to bring to the interviews their own Statement of Achievement based on a summary of their Profile records. Tutor A also brought to each session a draft of his Tutor's Review, prepared with reference to pupils' draft statements which he had already seen. He had given pupils photocopies of his draft Review statements some days earlier and had encouraged pupils to take them home to discuss with parents. In contrast, Tutor B, who was also a head of department, said he was not willing to give a similar amount of time to preparation and admitted that his main aim was to 'get the most accurate profile of the pupil in the least possible time'. Thus he intended to use the 20-25 minutes available to draft his Tutor's Review with the pupil present.

Both of these elements (i.e. pupil self-accounting and teacher-pupil discussion) are discussed in the previous and following sections respectively.

Our evidence allows us to go somewhat further, however, in identifying the essential differences between the old and new teacher-centred processes. First, statements and assessments of the RoA variety tend to reflect, to a much greater degree than do conventional teacher reports and assessments, the principle of positive description, and they tend moreover to be based upon more comprehensive evidence.

Evidence available to fifth year form tutors at Hydrangea School, ILEA, at the time they composed the School Statements (for the London Record of Achievement) and upon which they were able to draw included: formative Unit Credits in years 4 and 5 from departments which had taken part in that element of the project; summative Credits from all departments; pupils' Student Statements; and of course evidence from their own interaction with pupils, in many cases over the whole of the pupil's secondary school career. There are clear signs in the sample of School Statements collected for analysis that this evidence was heavily drawn upon.

Despite the tremendous investment of time and effort involved in producing around 25 statements which attempted to portray positively each pupil's individual achievements over five years, tutors applied themselves to the task with great commitment. The main problem identified by several fifth year tutors was the principle of positive-only reporting. Tutors generally subscribed wholeheartedly to this principle, and genuinely wished to produce

summative statements which reflected on their tutees' achievements in positive terms. However, a few felt that it was unfair to the majority of pupils, who had made a positive contribution of one kind or another to school life, to describe in positive terms pupils who had made sustained and determined efforts throughout their school careers to make life miserable for staff and pupils. The kind of language used in a few of the School Statements which were collected for analysis reveals the difficulty tutors had in framing their portrayal of some pupils in positive terms:

'He needs time to build up trust with peers and adults.'

'There are signs that his readiness to persevere is developing, particularly when he sees that he will directly benefit from this.'

'His strengths have been in art, craft and design and English Language, though his natural ability has tended to fuel him along rather than any great commitment to perseverance and industry.'

Many of the new forms of teacher-centred assessment are based on criterion-referenced frameworks, such as comment banks or graded/staged assessment. Whilst these new frameworks for teacher assessments have in a number of pilot schemes been closely associated with the RoA project as a whole, there is also a sense in which they could be (and in some contexts are being) developed entirely independently of the other features of RoA systems. For this reason we have decided not to address the issues arising from teachers' operation of new assessment frameworks in their full complexity here. Issues relating to the content of these frameworks and to the notions of achievement which underpin them are discussed above in 1.1.

On the other hand we would want briefly to convey the experience of teachers in our case study schools who have attempted to use the various new frameworks for assessment and to integrate those frameworks with other classroom processes. Although teachers have generally acknowledged that the frameworks are conceptually impressive in their comprehensive coverage of achievement in a particular subject area, and in their detailed descriptions of levels of achievement, they have nevertheless found the frameworks to be too elaborate, bewildering, time-consuming and unwieldy to be of regular practical utility in the classroom.

The pilot work for the Modern Languages G Component at Wisteria School, OCEA, never progressed sufficiently far to enable a detailed analysis of its validity. Staff faced considerable difficulties in their efforts to become fully conversant with the structure and meaning of the assessment framework, and this fundamental sense of frustration

and alienation, compounded by the effects of teachers' sanctions during the dispute, inadequate training of staff involved in the pilot, and the abiding sense that the assessment framework was largely provisional and subject to modification at any time, resulted in a state of limbo which characterised the pilot work throughout the two-year period. The inability of the department to implement the framework was not due to a 'validity' problem, i.e. to an inherent mismatch between the department's curriculum and pedagogy on the one hand and the principles underlying the framework on the other, but to the persistent judgement by members of staff that the package was technically flawed and unworkable.

1.2.3
Teacher-pupil discussion

The DES policy statement recognised that some form of teacher-pupil discussion would be needed in order to realise RoA principles:

> '. . . schools will need to set up internal arrangements for the compilation of records throughout a pupil's period of secondary education which will involve all teachers concerned and allow for appropriate discussion between teacher and pupil' (para 12).

Indeed the kinds of teacher-pupil discussions which have taken place as part of schools' RoA systems, together with the forms of pupil self-accounting described in 1.2.1 above, comprise the processes which are most often associated with RoAs and which distinguish RoA systems most clearly from other developments in the assessment, recording and reporting field.

(a) Purposes

All schools have incorporated some element of teacher-pupil discussion, although the purposes and contexts for these discussions have varied considerably. Certainly there has been no sense in case study schools that teacher-pupil discussion should necessarily be built into every element of an RoA system. In particular, it has been comparatively unusual for subject-specific recording to be conducted through formal discussion, although this has possibly been less a matter of choice than a response to limitations in resources, especially with respect to time and staffing.

For a number of inter-connected reasons, teacher-pupil discussions did not feature especially prominently within the system at Bluebell School. Most of the explanation for this relative absence of 'interviews' lies in the curriculum-orientated nature of the system. First, since most coursework profiles promoted self-assessment and descriptive judgements by both teachers and pupils, there was no need to negotiate the 'content' of a profile in the way that other systems based on ratings or selections from a comment bank have required. Second, most subject teachers in the school preferred to use written media to communicate with pupils and to engage in dialogue and review. One-to-one interviews were seen as (a) too time-consuming, (b) duplicating the function of the (written) profile, and (c) less reliable in tapping pupils' own thoughts and responses. Furthermore, written profiles were more consistent with the ethos of 'pupil ownership and independence' which the project aimed to foster. Third, the nature of subject teaching, which provided more contact time with pupils than tutorial work and which was perhaps more amenable to independent learning approaches in the classroom, was felt to lend itself more to brief and spontaneous one-to-one chats with pupils in the course of lessons, rather than the more formalised interviews which have characterised some systems. Fourth, since subject profiles dealt with aspects of achievement which were perhaps less confidential than those orientated towards personal qualities and out-of-school activities, staff felt less need to ensure privacy when discussing them with pupils. And finally, whereas pastoral work has no clearly defined cycles and so tends to require the more artificial interview to enable reviewing and recording, units of work in subject lessons presented clear points of completion, when reviewing and recording could be introduced naturally.

We certainly would not wish to suggest by this that processes of teacher-pupil discussion have not been incorporated into subject-specific recording elements. Indeed some of the most lively and balanced teacher-pupil discussions we have observed have taken place between subject teachers and their pupils, but these have tended to be spontaneous one-to-one chats with pupils in the course of lessons, rather than the more formalised interviews typically associated with tutorial-based discussions. It has also been on the subject side that we have observed the discussion serving diagnostic and target-setting functions.

Among the most common purposes for which the more formalised kinds of teacher-pupil discussions have been used are:

● To assist pupils with the process of continuous recording. In the early stages of the pilot, teachers (usually form tutors) used one-to-one discussions to help pupils to be comprehensive in the kinds of achievements they considered for inclusion in their diaries. This purpose has naturally become less prominent with the gradual discontinuation of regular recording described above in 1.2.1. Discussions continued to be used for this kind of purpose within RoA systems based on a more limited form of continuous recording such as pro-formas.

● Discussions have been used to serve the more general pastoral purpose of enabling teachers (usually form tutors) to get to know their pupils 'as people' by discussing their interests and activities outside the classroom. Although these kinds of discussions were envisaged to contribute eventually to the recording process (e.g. continuous recording and statement-writing), they did not necessarily result directly in a written product.

Discussions in the context of Guidance lessons at Campion Upper School, Suffolk, were of necessity brief and focused specifically on draft pupil self-accounts (in their formative Profile or their final Statement) or draft Tutor's Reviews. In those observed, the tutor discussed Profile contributions and made comments and suggestions about content and structure and occasionally corrected spellings. Specially arranged private one-to-one discussions were directly observed with six pupils in one tutor group in the fourth year. At this time there was no pressure to produce a summary statement so the time was used 'formatively' by the tutor to discuss pupils' participation in extra-curricular events, their out-of-school activities, sometimes their adjustments to new family circumstances, their reactions to the assessments made by subject teachers in connection with the twice yearly non-RPA Review, and their progress on GCSE coursework. In other words discussion was free-ranging and not strictly confined to the RPA headings. Indeed RPA was rarely mentioned and although pupils brought their Profile folders with them these were only directly referred to on two of six occasions. There was at this point no pressure for an end-product so the discussions were divergent and relaxed. The overall impression was that the tutor saw this as an opportunity to learn more about individuals in his tutor group and to give pupil's ideas about what they might further record. The value of this experience specifically in terms of the professional development of teachers was confirmed elsewhere. The school co-ordinator reported that a senior head of department of long standing had commented with surprise and obvious approval on the amount that he had been able to learn about pupils from his first one-to-one discussions with his tutor group. Indeed, the evidence suggests that it was this opportunity for tutors and pupils to explore personal qualities and achievements in the context of dialogue which was both the essence of the RPA system and the aspect which was most highly valued at Campion School. Although there were numerous grumbles about other aspects, notably the time taken for statement writing, there was evidence that some staff would regret the diminution of discussion of personal achievement which might follow from the replacement of the pastorally focused RPA by a more broadly-based RoA encompassing academic achievement.

● Discussions have been used to enable teachers to support pupils with the task of statement writing. Attention has often focused on a draft of the statement prepared by the pupil.

Records of Achievement

At Columbine Middle School, the second round of interviews between tutors and pupils in the first year of the pilot took place after half-term in the summer term, and were geared more or less exclusively towards the production of a negotiated statement by pupils. Pupils were expected to have prepared drafts of their statements in advance of the interview, and the meeting was then devoted mainly to discussing the draft and possible ways in which it might be changed and/or supplemented before being produced formally. For many pupils the preparation of the draft statement had been done in tutorial lessons with the help of their tutor anyway, in which case the interview was a more private and concerted follow-up.

Group interviews were the only interviews observed at Honeysuckle School, although provision was made for individual interviews. In two such interviews conducted by the same form tutor and involving a different group of three pupils on each occasion, the tutor presented the pupils with assessments of their practical and personal skills that had been made by staff anonymously. Pupils were asked to examine these assessments and ask questions. A short amount of time was devoted to a study of the assessments and pupils spent much of the interview writing their contributions for the summative document.

● Teachers have met their pupils in order to consult them over the content of teacher-written statements (see the illustration from Hawthorn Upper School in 1.2.2 above).

● Teachers (usually form tutors) have met their pupils in order to review progress across and beyond the curriculum over a fixed period (e.g. one term). Again, although these kinds of discussions were envisaged to contribute eventually to the recording process (e.g. continuous recording and statement-writing), they did not necessarily result directly in a written product.

● Teachers and pupils have met in order to compare notes in relation to some kind of assessment framework, either of subject-specific or personal achievement. Pupils' self-assessments have been compared with teachers' assessments, often with a view to reaching agreement.

The assessment of personal qualities formed an important part of the pastoral 'attitudes and relationships' document at Begonia School. The qualities appearing on the document: adaptable, interested in personal progress, organised, punctual, co-operative with classmates and adults (separated into two categories for fourth year pupils), acceptable appearance, sensitive, reliable, active, a leader, and attendance. With the exception of attendance, which was recorded as a figure, all qualities were assessed by both form tutors and pupils according to a four-point scale (always, usually, occasionally, rarely) and these assessments were discussed in reviews although both sets of assessments appeared on this document.

Considerable preparation went into the planned discussions between form tutors and girls at Magnolia Girls Grammar School. Both had produced assessments on grids and the former had collected and collated the assessments of all other staff teaching each pupil. The principal object of the discussion was therefore to produce an agreed statement based on these assessments. The evidence of five such interviews suggested that, in the limited time available (approximately 20 minutes) this agenda virtually crowded out any other. The imperative was to agree an assessment and little time remained to explore discrepancies in staff and pupil perceptions, analyse reasons for strengths and weaknesses, or discuss possible courses of future action. This is not to say that some of these things were not attempted; it was simply that the pressure of time was an obvious and powerful constraint.

A series of one-to-one discussions were observed at Hydrangea School, ILEA, between the head of the Modern Languages Faculty and half a dozen of her fourth year students of German. These were held at the end of a Unit, and were envisaged as a supplement to the exchange of information contained on the Unit Credit. Pupils were given an opportunity to discuss particular successes as well as particular problems they felt they had experienced in the context of the Unit which they had perhaps not been able to describe in the self-assessment portion of the Unit Credit.

As some of the above examples illustrate, on any one occasion a number of these purposes could provide the basis for discussion depending on the stage reached in the total RoA process in the school.

A few general points can be drawn from this wide range of purposes and contexts for teacher-pupil discussions. First, as has so often been acknowledged in this report, much of the work in this area has been carried out by form tutors. Recognition must be given, not only to the multiplicity of purposes for which their discussions with pupils are likely to be used, but to the increased needs for time and training stemming from this new and additional work. Second, in keeping with the general trend away from purely formative processes which we have already mentioned, it must be reported that, as pilot work progressed, teacher-pupil discussions came to be used more and more to serve summary purposes (e.g. geared toward interim statements) as well as, or instead of, the more exclusively formative ones listed above. This may well be at least in part a result of the awkwardness experienced by teachers and pupils when trying to conduct a discussion which lacked a clear and tangible focus. By this we do not necessarily mean that a specific summative purpose was needed; merely that a pupil's feelings, interests, values, social life and family life may have been at once too amorphous and too volatile to serve as a topic for discussion without a more distinct focus.

In the course of data collection we observed and tape-recorded dozens of RoA-focused teacher-pupil discussions and we have presented analyses of many of these in our individual case study reports. It is difficult within the limitations of this report to do justice to the many issues arising from those detailed examinations. We encourage colleagues with an interest in the nature of talk within RoA-focused teacher-pupil discussions to refer to our case study reports. Here we have drawn out a selection of issues arising from those reports which give a flavour of our analyses.

(b) Nature of teacher-pupil talk

Many of the questions which preoccupied us when investigating teacher-pupil talk within RoA discussions were long-term, research-type questions which were difficult to test within the terms of reference framing our evaluation. For example, we are interested in the relationship between the 'rhetoric of negotiation', which characterises the RoA movement, and the reality of interaction between teachers and pupils in actual RoA discussions. We are also interested in what might be called the 'language in education' issues surrounding RoA discussions: do teachers see such discussions as opportunities to foster a particular kind of oral communication competence in pupils, which might link up with the recent growth of development work in oracy, or are the discussions viewed exclusively as exchanges of information, opinion and judgement? A great deal more research needs to be done before we have satisfactory answers to these kinds of questions, but we hope that our work may at least begin to cast some light upon them.

Any examination of the quality of RoA-related discussions should be informed by the quite considerable research literature on the nature of teacher-pupil talk. One of the basic principles underpinning that literature is that the quality of talk between teachers and pupils is strongly dependent upon the **context** for the talk, and that this context includes such factors as:

● the physical and social setting,

● the implicit understandings which may have been built up between teacher and pupil over time,

● the way in which teacher and pupil understand the purpose of the task upon which the talk is focused,

● and the wider societal and professional structures and institutions which contribute to teachers' and pupils' conceptions of their roles.

Clearly one element of the context for the teacher-pupil discussions we have observed in RoA pilot schools is the specific RoA system within which the discussion is taking place. We have in our case study reports often remarked on the way in which the nature of the talk between teachers and pupils in an RoA interview or discussion appears to be significantly determined by the kind of RoA system which is being handled by the discussion. For example, teacher-pupil discussions taking place in the context of RoA systems which require specific judgements to be made according to stated criteria, categories or levels of achievement tend to be constrained by the limited range of terms in which 'agreement' between teacher and pupil can be expressed, despite the fact that there may be advantages of a different kind in having a clearly structured focus for the discussion. The talk within such discussions tends to concentrate on (a) gaining a clear understanding of the meaning of the terms framing the judgement and (b) trying to reach some kind of consensus within those terms of reference.

Many of the 'disagreements' observed between tutors and pupils at Daffodil Comprehensive School, Wales, in their interviews focusing on the 'Record of Personal and Social Development', appeared to stem as much from uncertainty over the precise meaning of many of the headings and statements as it did from genuine conflicts of viewpoint. Under 'Assertion and Resilience', for example, two alternative statements were available:

'Confident and assertive'; and
'Avoids drawing attention to self'.
Most pupils, and indeed tutors, attempted to discuss such groups of statements as if they were not only mutually exclusive but also hierarchical, i.e. that 'confident and assertive' was 'better' than 'avoids drawing attention to self', and moreover that it was not possible to posses both 'qualities'. However, it was sometimes difficult to sustain this principle.

'Disagreements', or polite differences of opinion, were dealt with with much embarrassment on both sides, especially where the tutor's assessment was less 'positive' than the pupil's. The solution to these differences was always initiated by the tutor, and tended either to be a provisional compromise pending a further interview, or the teacher's own assessment prevailing.

Teacher-pupil talk within systems which were based on more descriptive methods of accounting, and in which the purpose of teacher-pupil discussions was to garner, examine and present **evidence** in a comparatively open-ended and collaborative way (e.g. statement-writing) tended to be less trammelled and thus potentially more responsive to pupils' concerns, since the talk could focus on the pupil and on finding ways of portraying him or her rather than on the form of judgement specified by the assessment framework. However, what might be gained in terms of potential flexibility in the talk might be lost in terms of a structured focus for discussion.

Another aspect of the context for teacher-pupil discussions, which had a bearing on the quality of the talk within it, was the strength of the relationship between teacher and pupil. This may seem like such an obvious factor that it hardly warrants mention. Nevertheless, there are several tensions inherent in this aspect which deserve examination. One would predict that the quality of talk in a discussion between a teacher and pupil who already have a reasonably good rapport would be more relaxed and lively and less impeded by the pupil's recalcitrance.

The tutor at Campion Upper School whose discussions we observed, initiated and guided the interviews, and his input was somewhat greater than the input of the pupils with whom he talked, but on no single occasion did he talk for any great length of time. This meant that there was a genuine interaction and exchange of ideas. Several reasons may account for his not dominating in the way that many teachers appear to do: he was unfamiliar with one-to-one discussions and appeared diffident (this may of course have been researcher effect); his mode of questioning underplayed his authority role (e.g. 'I'm not sure I know the answer to this but I'm just interested to know'); he admitted to being slower than many of his colleagues which would account for the slow pace of the discussions and the frequent pauses which had the wholly positive effect of giving pupils space to think and to frame responses; he had been tutor to these pupils since they entered the school and expected to be so until the end of the fifth year therefore his identity as their pastoral tutor was well established. When interviewed subsequently, pupils reacted positively and said they felt that they had been able to say all they might have wished to say.

Maintaining psychological distance and authority seemed to matter less to one tutor at Hawthorn R.C. Upper School, perhaps because he was more junior in both age and status. His discussions were conversational with a fair amount of good-natured humour of the leg-pulling and tongue-in-cheek variety. Oral contributions from pupils almost equalled those of the tutor, in terms of quantity, and pupils frequently asked questions themselves, mostly seeking advice. It was evident that the tutor knew a lot about them as individuals: about their family backgrounds and their aspirations. These were often mentioned although the tutor was aware of the dangers of inappropriate disclosure which had been discussed fully in a preparatory INSET session. Pupils were clearly impressed by the draft Tutor's Reviews but sometimes thought they were over-flattering. The tutor made it clear that he expected document users to read between the lines but one pupil suggested that his statement would be more convincing if the good were put alongside the bad, for example: 'In the past J has . . ., but now he is . . .'. Another pupil persuaded the tutor to omit a sentence, which had a predictive quality, because he did not want to create an impression that he might not be able to live up to. Negotiation of accounts was therefore a genuine feature of this Tutor's discussions with pupils.

Tutors at Buttercup R.C. School, Lancashire, reported that it was interesting to find out about pupils they had little contact with and that it was difficult to know if a pupil was exaggerating their achievements if the tutor knew very little about them. For one tutor the focus of the review was a business-like one with the emphasis on getting a job done, i.e. getting the progress profile filled in, rather than on the development of a rapport between the tutor and pupil.
Following his first round of interviews with pupils a new reviewing tutor reported that he had felt under pressure to fill in the lengthy checklist when he would have preferred to use the interview time to get to know his reviewees, some of whom he had had little contact with.

There is therefore a certain irony in the fact that the majority of RoA discussions we have documented in our case study reports have been between pupils and form tutors who report that they do not know each other especially well. This irony is exemplified by the frequent 'getting to know you' purpose for which so many discussions have been used. In several of our case study schools pupils are allocated to a new form tutor every year or two — a practice which, while possibly sound in respect of some pastoral objectives, certainly militates against the building-up of the kind of rapport which ensures open and balanced talk in one-to-one discussions.

It is also somewhat ironic, and perhaps rather telling, that the teachers whom pupils tend to know best, i.e. their subject teachers, make little use of the characteristic one-to-one RoA 'interview'. It is instead their form tutor, whom they may see frequently but not typically in a context which engenders a sense of shared experience and understandings, who makes most use of the interview technique. The 'getting to know you' function of discussions between tutors and pupils may therefore be a legitimate and necessary one, but it should not be surprising if the talk within those discussions is more stilted and self-conscious than in discussions between teachers and pupils who already know each other well.

Although the personal tutor at Nasturtium Tertiary College, whose work was observed, clearly had a productive relationship with her tutees, there was some suggestion from elsewhere that students generally preferred to talk with their subject lecturers and course tutors about feelings and worries because they had more sustained contact with them; they often shared a particular perspective and many problems were indirectly course-related.

The other aspect of the context for teacher-pupil discussions which we have examined is the powerful determining influence on the quality of talk exerted by the traditional imbalance in power relations between teachers and pupils at school. The research literature on classroom talk consistently reveals the controlling and interrogational nature of teachers' questions, the asymmetrical speaking rights implicit in classroom discourse, teachers' low tolerance for silent thinking-time on the part of pupils when asked a question, pupils' strategies for getting the 'right' answer, and so on. It would be surprising indeed if these powerful patterns were miraculously absent from RoA discussions, simply because the nature of the task was slightly different from those to which teachers and pupils were accustomed. And, indeed, we found plenty of evidence that these phenomena also characterise pupil-teacher discussions with respect to RoA, although teachers were often aware of the problems and attempted countervailing tactics.

In all cases the largest oral contribution to those discussions observed at Rose School was made by the form tutor. Often this exceeded two-thirds of the total talk and pupils' contributions were often limited to single words or short sentences. However, this was not simply evidence of the difficulty tutors have in relinquishing their accustomed authority roles — although this is likely to be one explanation. Other factors undoubtedly had some influence. For example, in the first set of sessions observed, the first year tutor spent some considerable time explaining how interim summary reports would be derived from diary contributions. These explanations diverted time from dialogue and put the pupil in an essentially passive role. Secondly, target group pupils were relatively young and perhaps still somewhat in awe of the largely traditional school to which they had come, and therefore inclined to be reticent. Finally, on reading transcripts, one tutor remarked on the amount that she had said and attributed this to her nervousness at being tape-recorded for the first time (a research effect).

If the proportion of tutor-talk is evidence of the difficulty staff have in relinquishing authority roles, there was also countervailing evidence of a practice adopted by one tutor to underplay her status as an authority. On a number of occasions she presented herself to pupils as ignorant, or non-expert, but interested. Thus she asked one pupil to explain the process of constructing a model in CDT; she admitted to another that she did not like writing French when at school; to a third she admitted an inability to produce representational art; to a fourth she admitted her failure to understand his mathematics examination paper; and to a fifth she said she had never gained a mark as high as 88 per cent and that she was not very good at forward rolls in PE. She was an historian herself! All these personal disclosures, and others of a less self-deprecating kind, can be interpreted as a way of encouraging pupils to talk freely about themselves, secure in the knowledge that they were not being judged by an expert. We have no evidence to say whether this was a deliberate strategy, or whether it was effective in equalising the balance of power by suspending the taken-for-granted authority of the teacher/tutor role. What we can say is that pupils were satisfied with the process and outcome of their reviews: they felt they could say all that they might have wanted to say.

One very interesting piece of evidence emerging indirectly from our work is that teachers to whom we have sent transcripts of tape recordings we have made of their discussions with pupils have, almost without exception, been quite horrified to see just how much they talk and how little their pupils talk, how controlling their talk has been and how submissive and compliant their pupils' talk has been, and how quick they were to fill up silences. It would be an oversimplification to interpret this quality of imbalance as a conscious need to dominate, when in fact it probably stems from a lack of familiarity with that particular kind of formal one-to-one situation, the genuine difficulties they have met in trying to get pupils to talk, a falling back upon habitual patterns of interaction, and a lack of training opportunities to learn new attitudes and techniques.

Although a few schools have endeavoured to set aside special time for subject reviews, many teachers have experimented with ways of incorporating regular one-to-one review into the pattern of classroom activity.

(c) Finding time for discussion
(i) Subject discussion

At Jasmine High School staff involved in reviewing as part of the PE/expressive arts RoA scheme were able to use a floating teacher to free staff for discussions with pupils occurring mid-way through the course. This floating member of staff was only available during this particular course which created a problem for staff carrying out final reviews with pupils after the course had finished. On the occasions observed by the evaluator, staff combined classes in order to release one member of staff for reviews (although one admitted to leaving his pupils to 'run riot' whilst he carried out reviews in another room).

Finding time for discussions whilst coping with competing pressures for teaching time and classroom management was one of the biggest problems articulated by departments (English, PE, maths, CDT and science) at Rose School. Three possibilities for coping with this situation were suggested by groups or individuals: instituting an assessment week, using exercise books as a focus for teacher-pupil written dialogue, or making dialogue integral to a department's style of work. The last option was the ideal and the Head of Modern Languages was confident that she could make it work. However in incorporating discussion into the pattern of her own lessons she found that she had to reduce syllabus content and teaching pace. Her evaluation was that this had positive advantages because pupils learnt more thoroughly. Other staff found such radical change in the pattern of lessons more difficult to implement or even contemplate. Some continued to see discussion as additional to teaching and were reluctant to lose what they regarded as essential teaching time; others felt that it would be impossible to conduct one-to-one discussions in practical contexts surrounded by thirty other pupils needing help with potentially dangerous equipment. For these the idea of a written dialogue had more attraction although we have little evidence of practice.

This evidence suggests that the only feasible way of resolving difficulties, in finding special time for one-to-one discussion in the context of subject-specific achievement, is to change the pattern and organization of teaching and learning in the classroom. This would probably require increasing the opportunities for independent learning so that teachers have more time to discuss individual progress. It would also involve challenging the priority which teachers have traditionally placed upon whole-class teaching time.

In addition to requiring changes in the practices and attitudes of teachers, these suggestions would create a number of other general and specific problems. There is the problem of classroom management and control. In particular, there is a difficulty for teachers of practical subjects — e.g. in monitoring pupils in a classroom where there is potentially dangerous equipment. Moreover, pupils may not respond well to discussions with teachers about their achievement in the public forum of the classroom. Large classes would also militate against such changes. Finally, there is a problem for teachers who only see individual pupils infrequently, such as RE teachers who take a class for only one period per week.

As indicated in the example from Rose School above, written dialogue may represent a solution to the problems posed by these circumstances, though there is the risk that such dialogue might disadvantage still further those pupils with writing difficulties who may already be overwhelmed by the writing demands of school work.

Written dialogue may, however, undermine the personalised and individualised principles embodied in the RoA initiative, which may be better served by face-to-face communication and the opportunities for clarification it affords. Once again we emphasise that the purposes and contexts for teacher-pupil discussions must be carefully thought through.

(ii) Discussions with tutors

At several points in this section on the processes of recording we have acknowledged the disproportionate burden of work which has fallen on the shoulders of form tutors as a result of the RoA initiative. This is no less true in the area of discussions, since form tutors in most cases have the heaviest interviewing 'load' and the least contact time with pupils. Moreover, even in schools where tutorial lessons have been created, pastoral work does not present the kinds of natural points for discussion and review which are more characteristic of the academic curriculum. The original targets for frequency of interviews, in some cases as often as twice a term, have been found to have been over-ambitious, which is in part a response to the practical constraints on achieving such targets and in part a function of the move away from purely formative discussions. Nevertheless, most case study schools which did not have dedicated tutorial time built into the timetable at the start of the pilot have now either created such time or are moving in that direction, mainly to ease the pressure on tutors faced with the task of interviewing a whole class of pupils. Some of these moves towards dedicated tutorial time have taken the form of tutorial lessons, during which tutors have attempted to integrate one-to-one discussions with other classroom tasks/business, while there is also a trend toward providing tutors with a protected non-contact period each week to enable the withdrawal of pupils from their subject lessons on an indiviual basis.

At Sunflower High School reviews carried out during 1985/86 were organised in blocks, with supply cover being brought in to release staff. However, because of difficulties in finding supply teachers, tutors could not always be released. It was decided at a meeting of tutors that an additional timetabled period should be provided each week for reviews rather than having a concentrated block of time which some tutors reported was quite exhausting. It was also felt that a regular rolling programme could be more easily established. Form tutors involved in reviews with pupils were given an additional non-contact period per week for these discussions. The school co-ordinator asked staff to use a different non-contact period each week in order to avoid taking pupils from the same classes. There were reports from form tutors that this was sometimes unavoidable and that subject staff occasionally refused to release pupils for reviews on the grounds that this was inconvenient.

At Begonia School form tutors were released from lessons for reviews with pupils through the provision of supply cover — each receiving four 50-minute periods. There were complaints from tutors interviewed that this was insufficient. Some tutors had asked for and received additional time away from classes but others reported that they had used free periods, registration time or time after school for review-related activities. From September 1987, a 50-minute PSE period was created for third, fourth and fifth year form tutors. This period was described as a PSE period and incorporated Active Tutorial Work (ATW) and RoA related work including reviews. As far as subject reviews were concerned, it was only possible to organise academic reviews for those subjects taught in form groups for pupils in their third year although at the time of writing it was not clear if only the 1985/86 cohort of third year pupils had been involved in such discussions. Supply cover was used to release staff from lessons. In the long term, the school co-ordinator felt that subject-specific reviewing would have to be built into the academic curriculum.

Both of these models carry with them considerable implications, not only for staffing resources but for whole-school policies which ensure the co-operation and understanding of colleagues when tutors wish to withdraw pupils from lessons.

1.2.4

Teacher co-ordination of records

The last of the four processes of recording which we consider in this section relates to the ways in which teachers have taken responsibility for collecting, collating and aggregating RoA-relating assessments and accounts emanating from the first three processes. According to the 1984 statement of policy:

> As regards the summary document of record . . . the Secretaries of State see advantages in an arrangement whereby a single teacher takes co-ordinating responsibility (paragraph 37).

The issue of co-ordination of records was thus confined to the final summative stage at the beginning of the RoA pilot. Whilst we have collected evidence which relates to the issue of co-ordination of summary records, we also have evidence relating to issues of co-ordination of records at other stages as well. In particular, the general trend towards the production of interim records which serve some public reporting function (which was largely unanticipated in 1984 and which we discuss elsewhere in this report) has resulted in unanticipated demands on teachers to take on co-ordinating responsibilities at these interim points. In several of our case study schools, the increased use of formative records as a medium for reporting to parents, especially by academic subjects, has made the traditional overviewing role of form tutors more difficult.

The development of Unit Credits at Hydrangea School, ILEA, raised important issues for form tutors. In particular, form tutors found the lack of uniformity in the timing of Unit Credits being issued by departments frustrating. Although the previous reporting system had been widely disliked for its superficial and mechanistic portrayal of pupils, tutors had valued the way in which that system had kept them informed of pupils' progress at regular intervals. Since Unit Credits were issued by departments whenever a Unit was completed, it was difficult for tutors to gain an overall view of their pupils at critical times in the year (e.g. parents' evenings).

Records of Achievement

At interim and final summary stages, we have observed several ways in which teachers have co-ordinated the various contributions to pupils' summary RoAs. First, and most commonly, has been the largely administrative role of collecting and collating assessments and records emanating from various processes of recording, and ensuring that they go forward to the interim report or final folder, wallet or portfolio. Whilst this role has been time-consuming, and occasionally frustrated by the tendency of pupils and teacher colleagues to lose records or to miss deadlines for completing records, it has been conceptually unproblematic.

Form tutors at Rose School had the key co-ordinating role and were expected to pull together both personal and curricular records into summary documents at various points in a pupil's school career. Problems were anticipated if tutors took this to mean that they were expected to negotiate the total record with pupils in their one-to-one review sessions. For example, early in 1986 one member of the senior management team felt it would be inappropriate for form tutors to reinterpret records of subject achievement if this was outside their field of expertise and if they were unfamiliar with assessment criteria. To some extent this situation was avoided by ensuring that subject records were summarised, discussed, agreed and jointly signed by subject staff with students before they went to form staff (as was the case with reports to parents from July 1987). The task of co-ordination then became mostly one of collating and stapling.

Second, the overview statements usually compiled by a single teacher or tutor, some of which were described in section 1.2.2.(a) above, constitute a form of co-ordination, since these statements have generally been seen as a synthesis of reports and assessments contributed by a pupil and his or her teachers.

Evidence available to form tutors at Hydrangea School, ILEA, at the time they composed School Statements for each pupil (for the London Record of Achievement) and upon which they were able to draw included: formative Unit Credits in years 4 and 5 from departments which had taken part in that element of the project; summative Credits from all departments; pupils' Student Statements; and of course evidence from their own interaction with pupils, in many cases over the whole of the pupil's secondary school career. There are clear signs in the sample of School Statements collected for analysis that all of this evidence was heavily drawn upon. The task of writing each pupil's Student Statement was nevertheless a substantial undertaking, since it involved not only synthesising a diverse range of evidence but also couching the statement in exclusively positive terms.

As a variant of this second category, our case study reports include evidence that the single teacher with co-ordinating responsibility (usually the form tutor) has been required to aggregate, rationalise or summarise a range of diverse contributions to a pupil's summary RoA into an abbreviated form, usually within systems where this represented a surrogate for primary evidence.

Subject staff at Honeysuckle School made three sets of assessments of skills and personal qualities during the pupils' fourth and fifth years. Form tutors were required to collate these assessments for the summative document. The school co-ordinator remarked that the most recent assessments would be given preference over earlier ones but the aim was to give the pupils the best possible comment or comments. In drawing up the final profile he reported that the most frequently occurring comments were transferred. What was written on the final profile may have been an amalgamation of several comments which was permissible as long as the comments were not contradictory. The assessments for the summative document were checked by the school co-ordinator and year tutors to see if what was entered was a reflection of what the majority of staff had said about a particular pupil.

Form tutors seemed to utilise different procedures when carrying out the difficult collation exercise. One tutor commented that personal knowledge of a pupil was important in deciding which comments should be recorded on the final document, perhaps overriding assessments made by staff (although the school co-ordinator reported that these assessments would be taken into account when the summative document was prepared). Another told pupils that he selected the most common comment or comments from those indicated by staff on the comment banks used for both the assessment of personal skills and personal qualities.

Concern about this collation process was expressed by the head of science who felt that individual assessments, however carefully made, were lost during this process and that it was not worth spending a lot of time on the assessments and making them too exact because the assessments made of a particular skill or quality were amalgamated with those from other departments. This seemed to have important consequences for the regard of the scheme. This particular member of staff reported that his department did not take the scheme very seriously at all because of the lack of a specific science comment. However, he reported that staff and pupil attitudes changed for the better during the profiling period chiefly because comment banks improved in clarity, precision and user friendliness in successive drafts. Concern was also expressed by another member of staff who remarked that there may be a wide variation in comments from staff but that these are balanced out 'in favour' of pupils and that something that was pertinent about a pupil could be 'washed away' by form tutors because they were required to summarise.

Finally, our evidence includes an example of what might be called 'co-ordination by committee':

Daisy High School adopted a formal procedure for collating the various contributions to the summative 'Personality and Attitude' statement for fifth year pupils. For each tutor group, a staff conference was held to discuss and collate staff and pupil contributions to the 'Personality and Attitude' profile, and to agree the comments from the bank which were to go forward to the final summative statement for each pupil. These conferences were attended by the relevant form tutor, the head of year, the assistant head of year and the school RoA co-ordinator.

Each pupil was taken in turn. Most of the comments in the bank were clustered in twos and threes, and it was only possible to select one comment from each cluster. The meeting was told the tutor's and pupil's original selections from the cluster, and whether or not agreement had been reached in the tutor-pupil interview. The school co-ordinator then informed the group of the outcome of subject teachers' contributions for that pupil for that cluster of comments, where such contributions were available. Staff present at the meeting then contributed any information they may have had about the pupil, relevant to that cluster, especially if they felt that the pupil was being undervalued by staff or if they had first-hand knowledge to which other staff may not have had access. In most cases, consensus was quickly reached and the meeting moved on to the next cluster of comments. In some cases, however, there were conflicts between the various contributors which were less easy to resolve, and the meeting then tried to find evidence, either in the pupil's folder or from staff present at the meeting, which might have enabled a consensus to be reached. Usually in such cases, the benefit of the doubt was given to the pupil, and the less favourable of the comments in the cluster only selected if there was persuasive evidence available to support such a selection.

The staff conferences were exceedingly time-consuming — some lasting as long as three hours for a single class. Those staff involved identified this considerable investment of time as the major drawback to the procedure, and at the time of writing were considering alternative ways of collating and adjudicating over the contributions to this aspect of the summative RoA. At the same time, staff were reluctant to allow the outcome of the tutor-pupil interview to stand unmoderated.

This evidence highlights the fact that the burden of co-ordination, at both the interim-summary and final-summative stages, has tended to fall upon the shoulders of form tutors. Whilst in most cases this burden is largely administrative, there are also examples where tutors were expected to make a judgement about pupils' overall achievement, or to synthesise, rationalise and aggregate the judgements of others. This raises an issue about their competence to evaluate aspects of pupils' achievement about which they have little knowledge. They may moreover be unfamiliar with the criteria by which pupils have been assessed in subject-specific areas but be responsible for rationalising and summarising those assessments. One solution may be to develop RoA systems in which tutors merely collate subject-specific records which have already been discussed, agreed and summarised with the pupil by subject teachers. Their responsibility could then be limited to an administrative one, possibly also contributing to the personal element. Another solution might be to encourage tutors to observe pupils in subject lessons and to confer with colleagues about the achievements of individuals. This solution would naturally have very considerable resource implications. In any case, recognition must be given to the fact that form tutors are likely to occupy the pivotal position within virtually all RoA systems, and that this applies not only at the final summative stage but at the interim-summary stage as well.

1.2.5.

Overview

Many claims are made for the capacity of records of achievement to promote improvements in pupils' motivation, personal development and academic progress. There are two ways in which RoAs might bring about these improvements. First, the processes of recording aim to make explicit many aspects of teaching and learning which have traditionally been left unstated, including purposes, objectives, short-term targets, and so on. The passive and dependent attitudes of pupils towards their progress and development at school have generally been blamed for poor motivation and under-achievement. Armed with clearer notions of what teaching and learning are about, and a language with which to think and talk about those notions, it has been hoped that pupils will be able to take over more of the responsibility for their learning. Second, the processes of recording aim to provide pupils with regular, detailed, positive and individualised feedback on their progress and development; motivation would thus be boosted by pupils' greater sense of success and accomplishment.

We continue to believe that these assumptions are laudable ones, and we have some, although not a great deal, of evidence to support them. Moreover, we recognise that records of achievement have increased the motivation and professional development of many teachers. This is not a trivial point, since we can reasonably expect better motivated teachers to communicate their enthusiasm to their pupils. We also have enough evidence to support the conclusion that pupils in our case study schools have grown in self-awareness and in their ability to reflect more acutely on their academic progress and personal development. Whilst we cannot attribute these changes unequivocally to the RoA pilot work, they are nevertheless significant and gratifying.

At Rose School the strongest evidence for increases in pupil motivation and performance, as a direct result of RoA processes, was offered by the Head of Modern Languages. She claimed that since she had introduced a system of recording only positive achievement to her department the attitude of pupils to learning had been wholly positive. Moreover on conventional tests, used in previous years, the performance of first years had shown a substantial rise in overall standard. For example, first year pupils of all abilities had achieved marks of between 4 and 10, out of 10, for writing a letter in French under examination conditions. This had been assessed on GCSE criteria and many pupils achieved a standard which would have been awarded a CSE grade in the fifth year. In other words pupils were achieving something where they might not have been expected to previously. Another factor influencing this result was the undoubted enthusiasm and commitment to the RoA scheme which this head of department communicated to her pupils. She was determined to make it work, therefore it did. Observations of this teacher's lessons confirmed the interest and involvement of pupils which she had described.

In general terms however, it was difficult to elicit responses from pupils in interview which conclusively confirmed the generally favourable perception of staff. Nevertheless, most gave the impression that they thought the RoA (process and documents) 'a good idea' or useful and a few gave examples of changes in their behaviour (e.g. working harder in maths, or playing out less after school in order to catch up with diary work). Completed questionnaires returned from 11 parents (of pupils in one form), who said they had some knowledge of the RoA

project as it affected their child, provided a little more evidence. One question asked: Do you think that the Record of Achievement has had any influence on your child's attitude to learning in school? 6 answered 'No'; 4 answered 'Yes'. The four who answered in the affirmative, expanded on the nature of such influence, as follows:

'Improved links between child and tutor. Improved ability to talk to an adult in a sensible and constructive way.'

'I think he is more interested in what is going on and feels he has a lot of encouragement from the staff.'

'The ability to speak up and not be intimidated when unsure of facts.'

'She was very proud to have been selected as I was.' (This last might refer to her selection for the PRAISE close-focus group — possibly therefore a research effect.)

Fifth year pupils at Hawthorn R.C. Upper School, Suffolk, thought the Record of Pupil Achievement was basically a good idea. They especially valued the opportunity to discuss their own final statement and their Tutor's Review, and thought that it had contributed to their personal development, had changed their attitude to school in a general way, or had encouraged them to improve some aspects of their school work. For example:

Researcher: Thinking about yourself, thinking about your work in school, has it had any immediate value?

Pupil: Particularly on the characteristics I think. Because what I think I am, and what other people think I am, is different and certain things I've found I could improve on.

Like my social capability or whatever. I had to improve on that and maybe some of my work as well. It has made me realise what my thoughts are, so that I can amend them.

It needs to be borne in mind, however, that these responses were elicited at the time when pupils were engaged in final statement writing. It was the perception of tutors that pupils became more interested and better motivated towards RPA tasks when the end was in view. Similarly the influence of those pupils who had actually used the documents should not be under-estimated. One girl interviewed had already used her draft statement in a college interview and her estimate of value rested on the fact that the RPA had been asked for, read and pronounced good.

Pupils at Cornflower Special School, OCEA, made frequent mention of their experiences and achievements in maths and science, both in interview with me and in their P component recording. This seemed to indicate that the two G components being piloted by the school were having considerable beneficial effect. The science teacher also claimed that the beginnings of self-assessment using 'translated' OCEA

criteria had positively enhanced motivation. In maths, too, the teacher reported that pupils had responded well to writing their own reports on maths projects. Our observations of lessons would support the claim that pupils were enjoying the work and growing in self-confidence. In interview one pupil volunteered his perception that he had improved because he had watched the teacher marking and had seen where he had gone wrong and where he might put things right. However, whilst undoubtedly stimulated by the OCEA initiative, some of these changes might be more directly attributable to other, though related circumstances. For example, in this school, OCEA was accompanied by the introduction of a totally new subject-based curriculum. Thus pupils might have been reacting as much to the novelty of having subject lessons for the first time.

As we reported in both of our previous interim reports, we are prevented by methodological constraints from going further than these rather tentative statements. Apart from difficulties over arriving at an agreed definition of what constitutes motivation and personal development, the task of isolating cause and effect in the complex social settings in which RoA systems operate is extremely problematic. We would have liked to have been able to report on the effectiveness of the processes of target-setting and diagnosis in relation to pupils' motivation and attainment. We can certainly report that RoA processes have provided considerable scope for the setting of short-term targets. These targets may have been explicitly stated, realistically attainable and set by means of dialogue between teachers and pupils.

In the context of their end-of-unit reviews with their German teacher, fourth year pupils observed at Hydrangea School demonstrated considerable skill in identifying and discussing both strengths and weaknesses using a vocabulary of achievement specifically related to language learning.

Pupil 1: Well, I don't really quite understand all these cases and things like that. I don't understand when to use them. I understand how the words change but I don't understand when to use them and when not to use them.

Teacher: Right, then we need to look at that a bit more. But obviously you've been doing it OK in what's been asked of you so far, but in the long term, that's something we need to watch . . .
How did you find the speaking?

Pupil 1: I think speaking is my worst area.

Teacher: Why do you think that is? What's the particular thing about speaking that makes it difficult?

Pupil 1: I suppose remembering the gender of the words . . . When I'm writing I can get the words in the right order and things, because I see what I'm writing, but when I'm speaking I tend to get a bit confused and often get words in the wrong order.

Teacher: Although, so do Germans! So it's not that bad really . . .

Targets may, on the other hand, have been more implicit in the activity of personal recording, and sometimes served a more generalised disciplinary function.

An element of diagnosis and target-setting was implicit in the 'Personality and Attitude' comment bank at Daisy High School. The intention was that the fact of having personal qualities which the school valued explicitly on the agenda for recording and discussion would achieve a formative diagnostic function, i.e. that pupils would more consciously aim to improve those personal qualities which in interview were identified as 'needing improvement'. Indeed one member of staff likened the process, of self-assessment and discussion of certain personal qualities, to 'driving past a police car on the motorway': it serves as a reminder that school staff place importance on these qualities and are thinking about them in relation to individual pupils. This attitude arose in part from a concern expressed by many staff in the school over what was felt to be a rapid decline in common courtesy and self-discipline among many pupils. At the time of writing no clear evidence was available to suggest that a distinct improvement in pupils' conduct and attitude had yet been achieved.

Despite the various forms of target-setting and diagnosis which we have observed, the evaluation of their effectiveness has been problematic. For example, the failure of pupils to meet targets they set for themselves, or to improve in areas they had identified as weak (and we have a number of such examples), does not constitute unambiguous evidence that target-setting is ineffective. Inversely, examples of pupils succeeding in meeting targets cannot be attributed categorically to the influence of the target-setting process.

Researcher: Last year, there were a couple of examples, such as the presentation of your work, where you felt that there was room for improvement, and you were quite happy for that one to be ticked, rather than whatever the other one is, 'excellent'. Now, this time you seemed very pleased with the fact that you felt you'd improved the presentation of your work . . . You seem pretty proud about the standard of your work now. Well, I'm pleased about that, too. What do you think motivated you to improve the standard of your work?

Pupil: I don't know.

Researcher: Did it just seem to happen? I mean, you're a year older. People do grow up and change, and that's fair enough. I just wondered if you'd say that it was because it was identified on that Record Card last year that motivated you to improve?

Pupil: Probably . . .

Researcher: . . . what I'm trying to get at is how often this actually came into your mind

as something that you actually wanted to be 'better' the next time it came round. Did you think about it from one time to the next?

Pupil: To tell the truth, a couple of days after I forgot about it . . .

Researcher: . . . Only you can tell me how much it affected you.

Pupil: After we done it, I did think, 'I want to get 'excellent' next time – it'd be really good'. But I started work and I had to think about that. I just forgot from then, and it just carried on.

Researcher: But your work **has** improved, and you **have** got 'excellent' now, which is nice. . . . I just wonder why?

Pupil: I dunno. It's just something that happened, really. It just clicked.

(From case study report on Daffodil Comprehensive School)

These kinds of methodological dilemmas would, we think, plague any attempt to demonstrate the effectiveness of RoAs in this inherently difficult area of affective, cognitive and behavioural effects. Nevertheless, we would not wish to deny improvements in motivation simply because it is methodologically extremely difficult to attribute such changes to recording processes. Moreover, it is probably unreasonable to expect marked improvements in motivation in these early days which are characterised by a degree of experimentation and uncertainty. The positive effects may not be fully apparent until the process and its products have established their credibility, and the establishment of wider credibility may in itself enhance pupil motivation as much or more than the processes of recording.

1.3
Documents

In this section of our report we examine the format, presentation and use of the various forms of documentary records which were the tangible result of the recording processes described in the previous section. In addition to formative records and summative statements, we identify two less familiar categories: primary school summaries and interim summary statements.

We should point out however that our information on user response, especially with respect to summative statements, is very limited. We sent questionnaires to parents of our study sample of pupils in schools which had issued summary reports to parents, and/or final summative statements, by the time we concluded data gathering (Christmas 1987). We also attempted to trace users of summative statements in relation to these same samples of pupils. This latter proved to be extremely difficult and only the scantiest information was forthcoming. Our questionnaires to parents elicited a greater response but we hestitate to draw any general conclusions of a quantitative nature from them; instead we treated such responses purely as an additional element of case data. We recognise that user response is of the greatest interest but the scheduling of our study, concurrent with development within pilot schemes, meant that such information was in most cases simply not available to us at the time.

1.3.1

Primary School Summary

The DES policy document stated that RoA processes in the secondary phase should begin with a summary of achievement at primary school (paragraph 35). We have little evidence that case study schools attempted anything systematic in this area. In general any such summary had a very low priority and reflected the various starting points for projects within schools. Relatively few schools started with year one: preferring, or being directed, to start with older groups and extending the project downwards.

In the few examples available to us, the greatest benefit was perceived to be less in providing base-line data from which pupil progress during secondary education might be judged, as in disseminating RoA principles and practices to primary schools on which secondary schools might subsequently build. However, there was also an intention to extend primary school record-keeping in a way that would provide more complete information on pupils at the point of transfer.

Periwinkle School had very strong working links with its feeder primary schools. Staff exchanges took place and primary schools were working with Periwinkle School to provide a mutually agreed transfer record. There was also a common maths policy which evolved in the link between primary and secondary school. There was a primary classroom on the Periwinkle site where primary pupils could have experience of being on the secondary school site. The tutorial team at Periwinkle were concerned that pupils should be prepared to cope with the demands of secondary school.

Pupils coming in from the primary school were asked to bring in a statement of their expectations and a profile about themselves.

The transfer document involved the pupils in writing about what they were doing at school: their hobbies and interests and their expectations for their new secondary school. They were also asked to write a personal statement in which they described these activities and interests.

The maths department at Marigold High School appeared to be the only one that investigated the possibility of a summary of achievement from the feeder primary schools. The department felt that a lot of time was wasted at the beginning of the pupils' first year at Marigold finding out what stage pupils had reached – a problem exacerbated by the fact that pupils came from different schools, each teaching different maths schemes. Visits were made to two of the three feeder primary schools by the head of maths. At one of these schools, it was suggested that the pupils' last exercise book should accompany the pupils to Marigold, with a page at the front indicating which topics had been covered and with what success. There were also to be discussions at this feeder school about the possibility of a record of achievement issued at the end of this phase of the pupil's schooling. Both steps were regarded by the head of maths at Marigold as being useful in maintaining continuity in the educational process.

1.3.2

Formative records

Formative records are taken to refer to working documents which are used as a focus for teacher-pupil discussion. According to our evidence, they were considered to be formative in one or both of two senses: by contributing to the educational development of the pupil to the extent that they enshrined diagnosis and target-setting, and by contributing to the production of summary statements. During the course of our investigations our initial assumptions about what counted as formative, as opposed to summative, documents were somewhat undermined as we observed that many documents which were used for formative purposes were in fact summaries of, for instance, achievement over a period of time or a unit of work. This became particularly evident as we witnessed a move away from continuous recording (see 1.2.1(a)) towards periodic recording. The defining characteristic was therefore not so much the content or structure of such documents as their concept of audience and use. Indeed the same documents could serve different purposes at different times. The distinguishing feature was therefore their function: whether they were serving a formative function, as described above, or whether they were a means of reporting to pupils, parents or other potential users.

The character of such documents, in terms of number, content and structure, varied tremendously according to the scope of the RoA system in a given school and the nature of the recording processes (described in 1.1 and 1.2). The following illustrate something of the range although they do not cover all the variants in our data.

During the OCEA pilot phase (1985–87) at Cornflower Special School formative documents comprised:

For the P component. Pupils each owned a box file containing a P booklet for recording of achievement and experiences, a cassette tape for oral recording, photographs and various other written records. The OCEA assistant also kept a record book in which she noted pupils' achievements in various lessons which she had observed.

For the Science G component. A teacher-designed record sheet was used to record criteria satisfied in relation to specific activities. These were transferred to the OCEA-provided group record book. Both were held by the teacher. At the point when pupils began using their own translations of OCEA criteria, their self-assessments were discussed with the teacher and contributed to his records.

For the Maths G component. Pupils compiled booklets in which were recorded the 'modified' OCEA criteria achieved; a new page was added with every new achievement so that gaps in the record were not obvious. The bulls-eye form in which achievements were recorded was adapted from OCEA-provided record cards but made more attractive with humorous cartoons. The teacher also kept a record book for each child containing test sheets and project comment sheets. The latter were principally filled in by pupils. The inclusion of photographs of project work was also considered but the maths teacher left the school before this idea was developed.

By September 1987 a number of changes were evident. Tape recording had been abandoned because it had not been successful and box files for pupils' personal use had been replaced by ring binders in which the principal item was a report form for use after each lesson **by teachers**. These were kept in an open cupboard (pupils had unrestricted access to their own files as did their parents and other professionals) and on the door were pinned examples of acceptable and non-acceptable reports. The key requirements were that reports should be positive, based on evidence and be diagnostic. There was room enough on each report form for four to six separate lesson comments. The Head looked at each form when completed and might choose to add a comment of his own. There was an additional form for comments by non-teaching staff and parents, whose attention could be drawn to the files when they came into school for their children's medicals. However, whilst some dinner ladies, for instance, had chosen to record comments, the school continued to have difficulty in engaging the interest of parents. The general emphasis, however, had passed from personal recording by pupils to recording by adults, albeit ideally in discussion with pupils, and the formative documentation reflected this.

Records of Achievement

In the first phase of the development at Poppy County High School, Essex, the comment banks developed for use in pupil-teacher discussion at various stages in years 1 to 5 constituted the formative records. In terms of format these were made up of lists of 'has done' statements against which teachers and pupils signed their initials if they agreed the achievement. At first the total comment bank with personal, general and subject-specific statements was circulated around all staff from the RoA office during designated RoA weeks but later, when a decision had been taken to hold subject-specific comment banks within departments, this procedure applied only to banks relating to personal and cross-curricular achievement.

In Autumn 1987, when the RoA system was totally revised by a new school co-ordinator, comment banks were replaced by a Record Book designed to perform both formative and interim summary functions. These were held in secure cupboards in form bases and included summary sheets to be completed jointly by pupils and subject teachers in mainly free-prose format, together with a record of general competences to which any teacher might contribute, and a personal record and questionnaire (upper school only) to be filled in by pupils. A further sheet was provided for overall assessments by the pupil, the form teacher, the year head, the head of school and the parent.

The formative record at Hawthorn R.C. Upper School was the Profile compiled by pupils in tutor periods. Each Pupil Profile consisted of a set of printed, blue, ruled, A4 sheets on which pupils made prose entries under the three main RPA headings: personal qualities; experiences of school and work; out of school interests and activities. They could make as many entries as they wished but at intervals they were expected to discuss and jointly sign these with the tutor. Each pupil's set of blue sheets was kept together in a clear plastic folder and stored in a lockable filing cabinet in tutor rooms.

These examples illustrate the difficulty of distinguishing formative from summary documents in terms of their format and presentation alone. Some reflect substantial changes in continuous recording practice whilst the others are associated with attempts to summarise achievement at intervals or in relation to units of work. But the boundary between what might be called primary records (all routine records) and secondary records (summaries) is blurred and the notion of an exclusively formative document has significance only in relation to the purpose which it serves.

However, it is still pertinent to question, as we did in our interim evaluation report, whether for reasons of economy and coherence all forms of routine recording should be conceived as formative RoA documents and whether their form and content should be governed by any special criteria e.g. no negative comments, no tick-box checklists. On balance our evidence leads us to support the contention that all records should be conceptualised as part of RoA documentation in order to avoid perpetuating the idea that RoAs are a bolt-on extra. They need to be the umbrella for all recording processes if they are to have any value at all. However, we see little point in establishing strict criteria to govern the form and presentation of formative documents. The only criterion that really matters is that they should be able to serve the formative functions of teacher-pupil discussion: facilitation of diagnosis and target-setting, and the provision of a suitable data-base for the production of summary reports. In this context the use of marks, grades, checklists and comments on weaknesses could be quite acceptable provided that they are interpreted for the pupil and supported by evidence, and provided they support constructive dialogue about achievement. In keeping with the distinction between form and use, to which we drew attention above, it is less important that the records themselves are framed in positive **terms** than that they should be used in positive **ways**. This places a professional obligation upon teachers to use formative records in a dialogue with pupils aimed at celebration of achievement and a constructive approach to diagnosis and remediation in areas of weakness.

1.3.3
Interim summary records

Between formative records and final summative statements has emerged another category of documents which are variously called interim summaries, pre-summative statements, end-of-year statements or unit credits. These came to have an increasingly prominent role as pilot work progressed in case study schools. However, as indicated above, the distinction between interim summary records and the other two categories depended less on form and content than on a conception of function. In all cases they summarised experience and achievement over a unit of work or period of time, but unlike purely formative records, they were conceived as having a reporting function, usually to parents but also to form tutors or sixth form staff. They were also distinguished from final summative statements in that they retained a prominent formative dimension and were used for diagnosis and target-setting and/or as a data-base for final summative records. In other words, in contrast with the other two categories of records, they had a dual function, although it should be noted that in some cases (the London Record of Achievement in particular) these interim summary documents were included in a portfolio of evidence accompanying the final summative statement.

The following examples are chosen to illustrate the conceptions of audience and the contexts of use for interim summary records and also the variety of such documents in terms of content and format.

The main types of formative records developed by some faculties and departments at Hydrangea School, ILEA, were Unit Credits. Departments were encouraged to develop Units and Unit Credits along the lines set out in the report by the Hargreaves Committee *Improving Secondary Schools*, which advocated the re-structuring of the curriculum into clearly-defined units of work and the production of Unit Credits as the medium for assessing, recording and reporting achievement in the context of these Units.

Generally speaking, a **Unit** was a clearly-defined course of work, with clearly specified content, and attainable and understandable objectives. Moreover, departments were encouraged to specify objectives for each Unit which included examples of all four of the Aspects of Achievement described in the Hargreaves Report. In this way departments aimed to specify objectives which included e.g. factual recall of information, understanding of new concepts, acquisition of new practical, oral and applied skills, and the development of personal and social skills and positive attitudes.

Unit Credits were usually single sheets upon which the scope and objectives of the Unit were briefly described and upon which a pupil's achievements in relation to those objectives were recorded. Unit Credits typically contained four elements: (a) a brief description of the Unit content and objectives, (b) a space for the teacher's assessment, sometimes including test scores, (c) a space for the pupil's self-assessment, and (d) a space for parents' comments. The Unit Credit therefore served multiple functions, including a record of the pupil's achievement within the unit, a medium for teacher-pupil communication and a medium for regular reporting to parents.

An interim RoA was issued for first year pupils at Marigold High School towards the end of the summer term 1987. This document was an A4 booklet with the school crest on the front cover together with the pupil's name. The document contained:

The 'About Me' section — Described as a record of what the pupil had achieved and experienced over the year, consisting of total attendance and punctuality figures, the total number of merits awarded plus details of the categories merits were awarded for, details of pupils' within- and out-of-school activities, pupils' feelings about their experiences and achievements over the year, pupils' plans for the second year and a comment by the form tutor. The section was signed by the pupil, the form tutor and the head of year. The section was hand-written by pupils and some attached photographs of themselves.

Subject-specific RoAs — Some subjects such as English, Latin, science and German used comment banks with relevant comments underlined. Others such as CDT, music and geography used grids illustrating four levels of achievement ('with assistance', 'with little assistance', 'with minimum assistance' and 'without assistance'). Others such as drama and PE made free written comments under headings. Not all subject-specific schemes made reference to personal skills and qualities. The actual content of this section varied between subjects but all provided free written comments under headings.

Parental contribution – Parents were asked to provide information on pupils' achievements that the pupils themselves had not mentioned, e.g. membership of local clubs or skills/jobs they were good at. They were also asked to comment on the performance of their child during their first year at Marigold.

The interim RoA was issued for first year pupils instead of a traditional report and it was intended that all such statements would replace reports throughout the school. Although regarded as a document that belonged to the pupil, the interim statement was collected back in once parents had seen it so that it could be referred to subsequently. Only one copy of each was made. Marigold made the decision to keep the interim statement for a year and then give it to the pupil to keep. In practice this did not work out because staff found that they did not need to refer to the document and it was felt that the pupils may have had the greatest need to have access to the information contained within it. There was support for this view from one pupil who said that she could not remember the contents of her interim RoA and so was presumably in a poor position as far as trying to improve on the assessments was concerned. In future, interim statements may be kept by the pupils and if staff want to refer to anything they can ask pupils to bring them back into school.

At Geranium School the interim summative statement issued for first year pupils during the summer term 1987 was an A4 booklet containing a subject-specific RoA from each department, a sheet containing a section for a free written pupil comment and for parental comment, and a sheet for assessments of personal qualities made by form tutors together with lists of pupils' school-based, home-based and community-based activities. There was space at the bottom of this sheet for a comment by the year tutor. The format for each departmental RoA varied. The school co-ordinator commented on his reluctance to be prescriptive about the format of the RoAs since each department had to design a document to suit its own purposes.
The school retained a copy of the interim RoA which was issued as a report to parents.
As far as use of the interim statement was concerned, pupils reported that they found the document useful as a source of information about their strengths and weaknesses. Parents commented they they found the detail in the interim statement useful at parents' evenings because they did not have to spend most of the little time they had with staff attempting to unravel the precise meaning of terms such as 'satisfactory' that used to appear on traditional reports. Parents commented on the lack of a standard format but whether one format would have suited all departments was questioned. Some parents said that they liked the variety which made the document more interesting to read, although one thought that standardisation of grades, and so on, used throughout the document would have been helpful. It was generally felt that the RoA was easy to understand although it was difficult to extract information from some of the formats used. They found the language used accessible and welcomed the amount of detail on their children's achievements. Some parents said specifically that they liked the 'Pupil Comment' because it provided the pupils with an opportunity to comment.

One parent felt that this was the most important section in the document since this was a sure way of detecting if the child was happy at school and with life in general. However, exactly what this section was intended to achieve was questioned. It was also pointed out that there were some contradictions between the pupil comment and staff assessments.

In these examples, interim summary statements, developed under the aegis of the RoA project, appeared to be replacing or transforming traditional reports to parents. In many of our case study schools there has been a growing awareness of a need to rationalise and co-ordinate the multifarious recording and reporting responsibilities of teachers and schools and it is felt that ultimately little value will attach to the RoA initiative unless it subsumes existing forms of recording and reporting. (This issue is dealt with in detail in 1.4.) We endorse this view and judge that it is entirely appropriate for reports to parents to be conceived of as interim summary RoAs. However, we acknowledge that a considerable problem persists in relation to the nature of such documents; in particular a tension is created by the aspiration that interim summary statements should combine both formative and summary functions in relation to a range of audiences.

This issue has emerged forcefully on case study schools which have applied the principle of positive only reporting at the interim summary stage, on the assumption that weaknesses and targets for improvement are the proper subject for teacher-pupil discussion and there should be no need for anything of a negative, and potentially demotivating, kind to be indelibly recorded in **any** summary document. Our evidence suggests that this has evoked a strong and sometimes hostile reaction from some teachers and parents who have either seen this as an evasion of responsibility or simply dishonest.

Parental response to the first interim summaries that went home in the summer and autumn of 1986 from Rose School was neutral or negative. The reason for this was that teachers' union action, which persisted until December 1986 in this county, prevented the issue of subject reports. Pastoral statements of personal achievements and qualities were alone of little interest to parents who wanted to know what percentage marks and grades pupils were gaining in their subjects, and what their positions were in class. However, after the first issue of EMRAP-style reports in both pastoral and subject areas, parental response changed and the school's own evaluation indicated a high degree of satisfaction.

PRAISE questionnaires to parents asked no specific questions about reports as such and responses to questions asking for parents views about the best and worst aspects of their child's Record of Achievement were interpreted in a number of ways sometimes referring to processes or formative records. However, on the positive side, comments included:

'I think anything that encourages children, especially ones who will not reach top academic standards to put more effort into their work is to be applauded.'

'The child is able to contribute and express concern or delight regarding his progress.'

'Non-achievers can have their good points spelled out.'

On the negative side, concern was expressed about the lack of negative comment:

'Not honest enough. If someone is lazy the report should say so, as it should all other faults. If children are stupid, lazy, ill-mannered, ill-disciplined then the document should say so.'

'I don't like it at all. It makes the child's work appear **far** better than it is.'

'No bad points are mentioned – therefore often left uncorrected.'

The issue of lack of negative comment was also a major bone of contention with staff, some of whom felt that the positive-only rule should be reserved for summative documents. The school's guidelines for reports permitted weaknesses to be expressed as future targets but explicity ruled out negative comments on behaviour. This was unacceptable to some teachers who argued that parents would be misled if reports gave the impression that the achievements of some pupils were almost wholly positive.

Other case study institutions have not taken this line and have felt it to be quite appropriate to communicate areas of weakness to parents at this interim reporting stage.

An interesting issue arose at Nasturtium Tertiary College over the ownership of interim reports. Under the old FE college regime, reports had been addressed to students and written in the second person. However, from the time when the college opened its doors to students who would formerly have remained in schools, it was felt that the college had an obligation to report to parents and provide them with diagnostic and predictive information. For this reason reports included statements at levels 1 and 2 which indicated weakness and were, with the exception of the statements of out-of-college experience, written in the third person. Since students using the computer programme were expected to make their own records they had the unusual task of writing about themselves in the third person, although there was little evidence that this posed any great difficulty. More problematic was the loss of student control implied in the requirement that reports should be sent to parents. Whilst students did not contest the right of parents to see reports, students' representatives objected to the fact that on one occasion reports had been posted home, which implied less trust than they had experienced in their secondary schools. Staff explained that there were purely logistical reasons for adopting this procedure on this occasion but the incident served to reveal an underlying contradiction that is not easily resolved.

We would argue that these two approaches rest on different conceptions of who is party to the formative process. In the first approach the assumption is that formative discussion properly involves only teachers and pupils whereas the second implies, not only that parents have a right to information about under-achievement, but that they can be brought into the formative process in a partnership role. On balance we feel that the second approach is to be preferred, not only because we would support parents' right to know the whole picture during the course of their child's education, but because they are in a good position to follow up the formative dialogue between teachers and pupils in school by encouragement and support in the home. However, any suggestion that positive only recording need not apply strictly to interim summary records in no way diminishes the obligation on schools to be as constructive as possible in any comments on weakness. Indeed if parents are to fulfil a partnership role they will need guidance on ways in which their children might be encouraged to overcome difficulty.

At another level this debate about the function of interim summary records, and in particular the question of negative statements, is also about conceptions of ownership. The approach which regards formative discussion as confidential to teachers and pupils, and admits only positive comments, is perceived to be less likely to compromise the widely-held principle that pupils should have a sense of ownership of both the recording processes and the documents. With good reason this is considered vital to pupil motivation. Implicit in the partnership approach however is a different view in which the conception of pupil ownership is limited to the final summative document (this more closely accords with the principle stated in the DES policy statement, paragraph 40). Schools which favoured this latter approach acknowledged the motivational impact of pupils' perceptions of their control over process and product, but a degree of pupil control was aimed for in the **process of negotiating the content of records** rather than by giving pupils absolute ownership of documents. In our case study schools we have detected a move to the perhaps less idealistic but more pragmatic second approach, which corresponds to the partnership approach outlined in the previous paragraph. In our judgement such an approach is quite consistent with the principles set out in the DES policy statement and may hold the best hope for implementation nationally.

1.3.4
Summative records

By summative records we mean the final documents which become the property of pupils. Within pilot institutions these have mostly been issued at the end of the compulsory phase of schooling i.e. in the fifth year, although for pupils continuing in the sixth form some serious doubts were expressed as to the need for a pukka document at this point. Interim summary records were considered adequate for the purpose of transfer to the sixth form although it was thought that a summative statement ought to be available at the point at which pupils eventually left school. It was acknowledged that this implied a need to continue recording processes post-16.

Whilst supporting the view that RoA processes should continue post-16, and result in the issue of a further summative document at the point when the pupil eventually leaves school, we are persuaded by other evidence that the practical difficulties of deciding early in the fifth year who should receive pukka documents and who should not, are simply too great. Moreover, the distinction between 'haves' and 'have nots' is potentially divisive. Of course, the resource implications of issuing two summative documents to some pupils are substantial and will need to be recognised.

(a) Format and presentation

At the close of our data collection, summative documents had not been issued in all case study schools although in most cases at least one cohort was expected to receive them by the summer term 1988. Summative documents available to us before this date exhibited a considerable diversity in form, length, content and means of production. As with other forms of documentation, content varied according to the different treatment accorded to the aspects of achievement covered (see 1.1), the nature of the contribution of teachers, pupils and others to recording processes (see 1.2), and the relative emphasis given to different conceptions of audience and use. Format and structure however tended to be influenced by assumptions about the amount of material with which various groups of users could be expected to cope. In some cases, but by no means all, a degree of uniformity in this respect was encouraged or required by the LEA or consortium scheme. The following examples illustrate a range of permutations on these dimensions.

> The summative document at Hawthorn R.C. Upper School comprised a folded A3 sheet with designated spaces for a record of courses taken and examinations entered, a pupil statement summarising the formative profile under the same three headings (7 lines each) and signed by both pupil and tutor, and a Tutor's Review signed by the tutor alone. Latterly it had been presented in a glossy folder with the county crest and a protective plastic wallet.

> At Hydrangea School the summative London Record of Achievement documents were contained in an A4, triptych, red, plastic wallet and comprised: (a) a Student Statement (written by the student); (b) a School Statement (written by the form tutor); (c) samples of work selected by the pupil; (d) certificates and other evidence of achievement selected by the pupil such as formative Unit Credits from their subjects, references from employers, work experience reports, and awards from sporting or cultural organisations; and (e) summative credits from each department summarising

academic achievement over, say, a two-year course. Summative Credits typically contained (a) a brief course description, including content and objectives, and (b) space for the teacher's assessment, sometimes under headings describing aspects of achievement. A single sheet which explained the contents and purpose of the London Record of Achievement was included in the wallet, which displayed the school's name and logo and the signatures of the ILEA Education Officer, the school's headteacher and the chair of governors. This covering note explained that the London Record of Achievement was the property of the pupil, and that the portfolio had 'been designed in close consultation with major London businesses under the auspices of the London Education Business Partnership'.

The summative RoA wallet issued to fifth year pupils at Daffodil Comprehensive School contained:

- a certificate of work experience;

- a personal statement written by the pupil in the fifth year, known as the 'record of personal experience';

- a summative 'personal and social development' statement consisting of selected statements from a comment bank of 13 statements derived from the 'personal qualities' section of the 'record of Personal and Social Development'; and

- a 'response to subjects' derived from comment banks in English, Mathematics, Modern Languages, CDT and Home Economics.

All these were presented in a blue card, A5-sized wallet.

The summative document at Buttercup R.C. High School, first issued for fifth year pupils in 1987, consisted of a glossy cream-coloured A4 cardboard folder with the words 'Record of Achievement' on the front together with the county council name and red rose logo. The document was designed by the co-ordinator at Buttercup. The folder was intended to house documents other than the City and Guilds profile so the City and Guilds logo did not appear on the front cover. Nor was the school name on the front. This was in order to facilitate the spread of RoAs into other schools in the county, since it was argued that they would feel more able to use a scheme with the county stamp on it as opposed to a school name. The school's name appeared on a separate sheet of A4 paper inside the folder. This document also had space for the name and date of birth of the pupil, a brief explanation of the contents of the RoA (signed by the Chief Education Officer), the signature of the headmaster and the date. A second A4 document explained the preparation of the City and Guilds profile to the reader. This document, emblazoned with the City and Guilds name and logo, was signed by the pupil, the reviewing tutor and the City and Guilds monitor. The profile consisted of two A4 sheets of parchment paper. The first of these was entitled 'Profile of Achievement' and consisted of computer-produced statements of achievement plus examples entered under each of the fourteen cross-curricular headings. No indication of possible levels was given and any heading that did not have an assessment was printed out by the computer. The second page was entitled 'Further Information' and gave details of courses/exams entered for and details of the pupil's main activities and experiences. The most up-to-date progress profile formed the summative document.

The format for the summative document at Begonia School, EMRAP, was agreed by all the school co-ordinators involved in the EMRAP project. This comprised three elements: a personal statement, a curriculum statement, and details of the pupil's experiences and achievements. A sheet giving the pupil's personal details (date of birth, courses followed etc.) was also included. The document was printed on pre-headed flecked paper to prevent copying and issued in a standard EMRAP folder showing the name of the pupil and the school. The EMRAP logo appeared on the front of the folder.

The summative RoA prepared by fifth year pupils at Wisteria School contained a pupil-written record of experience in and out of school, a record of work experience, and a curriculum vitae prepared by the pupil using a computer programme written by two members of staff in the school.

These examples illustrate variations in length, format and presentation along a dimension ranging from the very synoptic to a more extended version comprising a kind of extended curriculum vitae, plus portfolio of supporting evidence. Nowhere did we encounter a final document of record consisting only of a portfolio of evidence, and we took this to indicate that users other than the pupils themselves were regarded, by schools, as the principal potential audience. However, in the light of our evidence of pupils' perceptions, that planning for nostalgia is an important function, we do not feel that this assumption can be accepted entirely uncritically.

All documents, whether or not they included a portfolio, nevertheless incorporated a summary record and it was possible to detect some emerging patterns in terms of the elements which comprised this. For example, all records incorporated some statement by the pupil, which provided a personal overview of interests, activities and achievements within and outside the school. All records also included some statement about courses followed although this element received a wide range of treatment and accounted for substantial differences in the length of summary records. The curriculum or course record, as it was often called, could be simply a list of subjects taken and examinations entered, or a one-page statement relating to the whole curriculum often with a few lines devoted to each subject or course. Alternatively it could consist of a one-page summary for each subject or course taken by the pupil. Thus the curriculum/course record could take up one page in the summative document or as many pages as the number of courses followed.

In addition to these two records which formed the basis of all summative RoAs, some case study schools also included:

- a tutors' review or school statement which was usually, though not always, discussed and agreed with pupils;

- a cross-curricular statement relating to personal qualities;

- personal and social skills or general competencies;

- a portfolio of evidence including perhaps interim records and examples of work; and

- a parents' contribution.

Most schools made provision, by means of a pocket in the document, for the inclusion of examination certificates when these became available.

When all these elements were collated the resulting document, or collection of documents, could be as short as four sides of A4 or much longer. The routine contacts which schools had with employers and managing agents led many schools to the view that a core summative record of from 4 to 6 sides of A4 was the optimum and that the pupils' own statement and a school (or tutor's) statement giving a general overview of achievement were the elements likely to be most highly valued. It was felt that employers would be less inclined to read detailed course records unless they had a particular interest in achievement within a subject area. Where detailed course records had been developed a folder with loose-leaf inserts was therefore considered helpful because it enabled pupils to select from the total record those elements which would be of particular interest to a particular user.

The issue of whether the various elements of the summative document should be hand-written, typed or word-processed was often raised but no consensus emerged. Views on the preference of employers for hand-written or typed pupil statements were more-or-less equally divided although typed school statements and other curricular records were generally preferred. Schools acknowledged however that typed or printed records required some level of clerical support which could not easily be met within their existing resources.

Another issue with substantial resource implications concerned the standard of presentation e.g. whether records should be produced on two-tone printed sheets and enclosed in glossy folders, or whether they could be far less polished and elaborate. During the course of the pilot phase a consensus emerged which endorsed the view that summative documents should look impressive if pupils, parents and other users were to endow them with value. Moreover, it was increasingly felt that special occasions when summative documents were formally presented to pupils also enhanced their importance.

At Daisy High School summative documents were presented to fifth year pupils at a formal ceremony just before the end of the Spring term 1987. The ceremony was attended by the majority of fifth year pupils, some parents and other local friends of the school. Held in the school hall, with members of senior management, LEA staff, civic dignitaries and the national evaluator seated on the stage, the event was seen as an opportunity to convey high status to the RoA and to commemorate the first full year of the project. Each pupil's name was read out, and they were presented with their RoA wallets by their tutor, the Acting Headteacher and the Chairman of the School Governors.

(b) Transition from formative and interim summary records

We turn now from questions concerning the physical structure and appearance of documents to questions concerning the nature and content of statements. The issues raised here overlap with some of those already addressed in 1.1 and 1.2 and illustrate the essential interconnectedness of the dimensions of the initiative we were seeking to understand.

In our interim evaluation report (1987) we identified an important issue concerning the relationship between summative statements and formative recording in terms of both process and product.

With regard to process, there was an issue about whether the principle of teacher–pupil discussion at the formative stage was carried forward to the summative stage or whether it was explicitly or implicitly contradicted if any record was written and signed by the teacher alone. In other words did the summative document validly represent the process that went into its production?

With regard to the products of recording processes, issues of validity were also raised. There was, for instance, the question of how summative statements were derived from formative records and interim summaries. Did they represent a selection from statements in formative and interim records or were they different in kind, in that they were in some way synoptic?

Our evidence indicates that case study schools approached the task in a variety of ways.

The summative record at Poppy County High School consisted of approximately 35 statements selected jointly by teachers and pupils from formative comment banks, plus examination certificates when these become available. The wording of the statements, selected from the four 4th and 5th year formative comment banks, remained unchanged. Statements were supposed to be selected to represent best achievement, but pupils sometimes selected statements which were less than their best achievements because they did not fully understand the hierarchical nature of statements within subject areas.

As with the interim summary statements derived by the same procedure but issued to parents of lower school pupils, the reaction of parents in July 1987 was almost wholly negative for many of the same reasons. This was considered disastrous because many parents and pupils were clearly disillusioned with the whole RoA idea which had been so heavily promoted. (It was not possible to get responses from employers etc. who might have used such documents.) As had been the case with interim summaries, a decision was immediately taken to abandon the comment bank approach to the production of summative statements and to institute a system very much more in line with developments at county level. In any case the LEA team had made it clear that the time for private enterprise was now over and a standard format would be mandatory. Thus the new school co-ordinator proposed that the information in pupil's Record Books should be summarised and typed onto the yellow and grey sheets printed by the county, using the secretarial support provided from the ESG grant. Potentially this was not too difficult a task because the Record Books were themselves summaries and the format for recording i.e. subject achievement, cross-curricular achievement and personal achievement, was roughly equivalent in both documents. The first issue of summative statements along these lines was not scheduled until summer 1988, so there was no opportunity to investigate user response.

At Bluebell School not all members of staff supported the idea of issuing summative subject RoAs, and even some of those members of staff who supported the principle objected to the decision to issue summative RoAs in 1987. Objections centred first of all on the conceptual and practical difficulties which departments encountered in moving from formative coursework profiles, which were geared to a limited and specific unit of work and to communication between teacher and pupil, to summative subject profiles, which were based on the principle that pupils' achievement over two years' work could be expressed on a single page and communicated to 'outsiders'. To some members of staff the two concepts seemed wholly incompatible; the only valid way of summarising achievement over an extended period was felt to be the presentation of all the coursework profiles which had been completed. Indeed, some departments actually attached a coursework profile to the summative profile they prepared on some students. The other objection was a more practical one, i.e. the lack of time available to form tutors to carry out the necessary discussions with pupils, collation of subject contributions and the preparation of pupil statements.

At Hawthorn RC School, pupils filled in the record of courses taken and examinations entered and used the material in their Profiles (the blue sheets) to produce a seven-line summary under each of the three headings. The extent to which these were genuine summaries varied with the skill of the pupil to precis; sometimes they more or less replicated the most recent blue sheet entry. Only 21 lines were allowed in total and there was a general feeling that this was too little although some pupils had difficulty in finding much to say about their personal qualities because there was a tendency simply to list them.

Despite this variety, most case study schools regarded it as especially important to discuss and agree the content of the final document with pupils, thus carrying forward the process of teacher-pupil discussion into the final stage. Furthermore, the balance of opinion was that summary statements should indeed attempt to **summarise** achievement over the period covered by the record. There was some anxiety about the reductionism that this implied but a genuine attempt to summarise was regarded as the only practicable and defensible way to avoid the overwhelming detail of the pure portfolio approach, or the atomism inherent in mere selections of comments from formative and interim summary records. This latter problem was particularly acute in comment bank approaches which did not have the capacity to summarise achievement at the transition from the formative to the summative stage. It is possible to conceive of separate comment banks for formative and final summative records — the latter containing comments which are more synoptic but fewer in number.

At Daffodil Comprehensive School the Upper School (years four and five) version of the 'Record of Personal and Social Development' consisted of 17 categories which were itemised on one side of the single A4 card: attendance, reasons for absence, punctuality, relationships with other pupils, relationships with adults, appearance, school activities, courtesy, consideration, co-operation, reliability, responsibility, self-discipline, assertion and resilience, oral communication, presentation of work, health as observed. A small space was provided for 'other relevant information'. Each category had two, three or four statements alongside it, e.g. 'excellent', 'room for improvement', 'cause for concern' and so on. These statements were envisaged as mutually exclusive, if not strictly hierarchical, and it was only possible to select one of the options at any one review (although half-way compromises were sometimes reached between some pupils and tutors). The grid was designed to enable five reviews during years four and five: three reviews in year four and two in year five.

The **summative** comment bank for Personal and Social Development was derived directly from the formative Record of Personal and Social Development, and was designed to enable the computer-aided production of a short list of positive statements relating to personal qualities at the summative stage. It consisted of 13 statements based on each of the first 13 categories of personal qualities on the Record Card. Form tutors selected those statements which they judged applied to each pupil in the spring term of the fifth year. Since each of the 13 categories generated only one comment, pupils at the summative stage were effectively deemed to have possessed a quality to an acceptable standard or not, rather than given a rating within the category.

On the whole, however, free prose formats seemed better able to portray the achievement of the pupil in a holistic way which could preserve individuality (comment bank approaches tended to create stereotypical images rather than differentiate among pupils), although the act of summarising demanded considerable skill on the part of teachers. Therefore, in endorsing the view that the task of summarising formative and interim records is an important aspect of the production of summative documents, we would also wish to draw readers' attention to the implications for the in-service education and training of teachers (see also 2.5). Summarising is a skill and many teachers will need in-service support before they feel confident in statement writing.

A particular problem which some of our case study schools have addressed, but which also needs further attention, concerns the question of how best to summarise achievement briefly whilst at the same time giving sufficient information about the content and context of achievement. Summative statements derived from comment banks have the advantage of brevity, whilst providing some breadth, because statements are often of the 'can do' or 'has done' type and give little or no detail of the specific context of achievement. On the other hand, evidence is easier to include in free prose summaries, which may nevertheless be lacking in terms of breadth because the prompts of the comment bank are missing. An appropriate balance between the need for reasonable brevity and the need for sufficient evidence and coverage is difficult to achieve and some decisions have inevitably to be made about priorities. One pattern of treatment that has emerged in some institutions is for curricular/course records to incorporate a brief statement of course coverage with summaries of broad areas of achievement supported by brief evidence of specific achievement within these areas.

This seems to us to be a very satisfactory way to approach the problem, although this amount of detail cannot be encompassed within a one-page whole curriculum overview; it requires at least half a page or a whole page for each course taken and favours the more extended kind of summative documentation. However, if schools adopt this general approach to the provision of information about the context of achievement then the question of how reference might be made to an explanation of coverage and grades in public examinations, as indicated in the DES policy statement (paragraph 23), ceases to be an issue. Course descriptions would be incorporated into the summative records for courses and results and/or certificates in public examinations simply have to be added when they become available.

A further general issue, one that was raised in relation to formative records and interim reports, is the crucial question of negative comments. The DES policy statement made it quite clear that summative documents should supply positive statements only (paragraph 13): a principle with which it is difficult to argue in view of the fundamental intention to enhance motivation by the celebration of success. This did not however lessen the tensions surrounding implementation. First there was the problem which schools perceived in meeting the expectations of users, especially employers.

(c) Negative comments and the relationship of summative RoAs to confidential references

Employers and other likely users of summative documents issued at Nasturtium Tertiary College in 1987 proved difficult to trace. However, one anecdote was instructive with regard to the policy of including only positive statements. A YTS managing agent had accused the college of misrepresentation in relation to a particular student. He said that he had taken this lad on the basis of a positive RoA but found him to be dishonest and disruptive. It emerged that the student had been excluded from college and forbidden to appear on the premises but the RoA principle prevented anything of this being communicated. Any credibility that RoAs might have had with this YTS agent was felt to have been lost.

An in-service session on final statement writing, held at Campion Upper School generated heated debate over the issue of positive-only statements. Tutors were not satisfied with the argument that employers should be left to read between the lines if they sought negative information. They also thought that they would be failing in their responsibility to parents and employers if statements created misleadingly high expectations. If this happened they thought the credibility of the whole exercise would be called into question. Indeed, on another occasion, one tutor quoted another as saying: 'There are lies, damn lies and there are RoAs'. The headteacher pointed out that this reaction illustrated a misunderstanding of the philosophy behind RPA and therefore the difficulty of communicating the idea that RPAs were designed to promote motivation and personal development. In his view, this would not be achieved by the inclusion of negative comments but if employers and others had any need for such information, it could be obtained through other channels such as confidential references.

Records of Achievement

At Honeysuckle School the summative document was described as 'belonging' to the pupils – that is, they could do with it what they wished. However, local employers had come to expect pupils to bring their profiles along to interviews, and it was regarded as suspicious if a pupil did not produce it. An additional problem affecting the use made of the Clwyd RoA was that pupils had conceptions of 'good' and 'bad' profiles. One pupil reported that if he received a 'bad' profile he would 'lose' it. The inclusion of negative comments such as the following one illustrates the dilemma faced by some pupils on this issue:

'X's dislike of school and her negative attitude towards authority has presented us with a number of problems especially in her fourth and fifth years. Her refusal to accept encouragement and advice from the staff together with her poor level of school attendance has not enabled her to gain a great deal from life in school. Her literacy and numerical skills suggest that she has satisfactory grasp of basic arithmetic and English. X appears to have a variety of hobbies outside school including art, swimming and ice skating. It is hoped that she will find something to her satisfaction in the future and we wish her well for wherever that takes her.'

(It should be borne in mind that this pupil was reportedly somewhat atypical and had given the school considerable trouble over her final two years.)

Scrutiny of the comment banks in use at Honeysuckle reveals the negative nature of some of these, e.g. 'X displays little interest in becoming involved in creative activity and work lacks (a) originality (b) imagination (c) enthusiasm'; 'X makes little effort to present practical work neatly'; and 'X does not show much inclination to spend time in organising his/her work effectively'. It is nevertheless interesting that both staff and pupils felt that negative comments had a place within the profile since the final document should be an accurate picture of the pupil, and since, to paraphrase one member of staff, even if it is damning at least it is truthful.

Reports were available from heads of fourth and fifth year regarding use made of profiles by local employers. One local firm was aware that the school issued profiles and always asked pupils to bring them along to interviews, but nevertheless still asked the school for a reference. It was reported that the personnel officer phoned the school to see if there was 'anything he needed to know' about a particular pupil.

Our evidence suggests that many schools had high hopes that employers would accept summative records in place of confidential references. Indeed teachers were often reluctant to engage in the elaborate task of producing final documents if they were subsequently going to be asked to write additional confidential references. Although doubts about the appropriateness and efficiency of creating a dual system remain, we now detect a degree of acceptance for the idea that additional confidential references should be provided by the school if they are requested. Nevertheless, case study schools propose to encourage employers to regard summative RoAs as a principal source of information, on the understanding that they record only positive achievement, and the degree to which they may be expected to differentiate among pupils is taken as limited. For this reason, the summative record is now widely regarded within our case study schools as a combination of an open testimonial and extended curriculum vitae. In our judgement this is an acceptable position to take and it is probably less important for schools to give employers exactly what they think they want (i.e. the 'whole truth') than that the precise nature and status of the documents are clearly communicated to potential users so that they can interpret what they see. In other words they should know that summative documents contain only positive statements, and what this implies, so that misleadingly high expectations can be avoided. (The illustration from Nasturtium College is a reminder that this will not always be possible.) In general, experience in case study schools suggests that this may be sufficient because comparatively few employers and trainers call for confidential references at the stage when pupils are seeking their first job or training placement.

The issue of negative comments is unlikely to be resolved, however, simply by an extensive publicity campaign aimed at employers. Our evidence continues to indicate that positive reporting can, and will be, subverted at school-level unless teachers and pupils have a thorough understanding of, and commitment to, the principle. The illustration below is by no means unique.

Although most of the Summative Credits available for analysis from Hydrangea School were written with a great deal of care, and were moreover detailed, individualised and positive, a content analysis of some Summative Credits revealed statements which were explicitly or implicitly negative, and in this sense perhaps more formatively-orientated. For example:

'J . . . is an infuriating pupil because he does have ability in Chemistry. Unfortunately he is all to [sic] ready to sacrifice this and be content with sloppy work.'

'W . . . needs to shown [sic] more enthusiasm for all aspects of the course, particularly problem solving which he finds difficult.'

'**Decision making:** With guidance. **Perseverance with difficulties:** Could show more interest in lessons. **Initiative:** Could contribute more in lessons. **Regular completion of homework:** Fair. W . . . works far below his potential. Most of his work is incomplete.'

'A thoroughly disruptive member of the group. V . . . has gained little from attending this course. He was never prepared to listen, accept constructive criticism or work as a member of a team.'

The comments of some members of staff suggested that they decided to comply with the principle of positive-only reporting on a superficial level only:

'J . . . when careful, can make accurate measurements and observations.'

'A . . . attended more than two lessons.'

'W . . . at times, put some effort into his work.'

'Would that all his lively contributions to the lesson were mathematical.'

Indeed with some pupils, staff clearly felt that they had insufficient evidence to report anything on a Summative Credit, positive or otherwise:

'I am unable to comment as A . . . has not attended regularly.'

'V . . . has taken little part in lessons making the recording of achievement impossible.'

This difficulty experienced by staff in writing positive comments about pupils who were disruptive and/or irregular attenders echoes the complaint of some form tutors, that it was unfair, to the majority of pupils who had made a positive contribution of one kind or another to school life, to describe in positive terms pupils who had made sustained and determined efforts throughout their school careers to make life miserable for staff and pupils.

Some teachers in our case study schools therefore remain unconvinced on this point and any national guidelines will have to take account of similar doubts in those schools as yet untouched by records of achievement. As in relation to many of the issues raised in this report, it is impossible to legislate to secure adherence to the 'positive only' rule. The limits of tolerance in interpreting such a guideline (e.g. whether to admit statements of qualified achievement) will have to be established at local level: within the school and in the context of in-service education and training.

(d) Ownership and use

The ownership of the summative document was essentially uncontentious. The DES statement (paragraph 40) that summative documents should be the property of pupils was generally taken as axiomatic within our case study schools. However, our evidence suggested that most schools proposed to retain a master copy in school files or on computer disk, at least for a limited period, but wider distribution was subject to the explicit permission of pupils. Schools expected to use these copies as a basis for formulating confidential references. This would seem to be to the advantage of pupils because, as teachers pointed out, such records were often the only source of positive information about a pupil. In the past, school records frequently contained only negative information: absence notes, court reports, memoranda about illnesses, letters home concerning problems etc.

As we pointed out in the introduction to this section on documents, our data on user response were very limited. We developed one questionnaire for parents and another for employers, trainers, and FE/HE tutors, who we were able to identify as users of summative documents from our close-focus samples of pupils. However, we were not able to gather a great deal of relevant information, partly because summative documents were only just beginning to be issued in many schools, and partly because it was difficult to trace those who had used such documents. However what data we had from parents indicated a positive response to clarity of presentation, to detail and to the intention to report positively. Where opinion differed it was largely in relation to those statements prepared in free prose format as opposed to those generated from comment banks. In relation to the former, parents appreciated the attempt by schools to give a rounded picture in which they could recognise the individuality of their child.

At Hawthorn R.C. Upper School, Suffolk, questionnaires were sent to external 'users' of summative RPAs issued to the group of pupils chosen by PRAISE for close focus study during the pilot period. In fact this amounted to only three of the six fifth year pupils in 1986–87 because the others transferred to the school's own sixth form. No replies were received. Responses to a questionnaire to parents of a whole tutor group sent approximately six months after the issue of summative RPAs was only a little more encouraging. Of 22 questionnaires sent, 9 were returned. Of these nine, 5 had no knowledge of the RPA; indeed, only three said they had seen their child's RPA document. However, these three – all parents of girls – were entirely satisfied with what they had seen and desired no changes or improvements. They volunteered that they found it comprehensive and attractive and they valued the attempt to give an overall picture of the pupil's achievements. Referring to the pupil's statement of personal qualities and the tutor's review, one parent wrote: 'Most children find it difficult to examine themselves objectively – this is a good exercise. The tutor's review is interesting as the tutor has a close personal relationship with the child and their opinion is valuable.'

On the other hand we had some evidence of a negative reaction to summative statements generated from comment banks, which, although detailed and clear, appeared impersonal and gave little sense of the wholeness, uniqueness or context of an individual's achievement.

At Honeysuckle School a training manager from a local company responded to the school's request for comments on the RoAs by saying that the comments were often difficult to interpret and suggested that a list of comment banks in use would be useful. This training manager also wanted to see a figure for attendance rather than a comment, and more detail of pupils' interests and achievements.

The difference in reactions to the content, form and tone of these two forms of summary document is very significant and leads us to the conclusion that comment bank approaches have little to recommend them, except in the formative stages of the recording process, unless some means can be found to develop systems which are capable of greater subtlety.

The very limited information we have of use by employers, trainers and FE institutions suggests that summative records were found most helpful in interview. They provided a focus for discussion and the pupils' personal statement was especially valued for the starting point it gave to interviewers who wished to encourage youngsters to talk about themselves.

At Cornflower Special School the evaluator interviewed three of the four pupils who left school in 1987, after their respective interviews for places in FE. (Employment is no longer a real possibility for pupils wishing to leave at 16.) All had taken to their interviews a copy of their personal statement — in draft form because the final documents came too late. According to pupils, no interviewer **asked** to see the OCEA statements although in two cases they expressed considerable interest when pupils **offered** them. The third pupil forgot to mention that he had brought his statement with him so it was not used. One pupil was very positive about it and felt it helped him to talk about himself in interview. The others still saw the summative document in a mainly formative light ('It's made me think about things what normally I won't think about') and were uncertain (through lack of evidence) about its value to their future tutors or employers.

Questionnaire responses were received from the FE tutor who had interviewed two of these pupils. In both cases he claimed that he had seen the Record at interview (although this contradicted what one of the pupils had said) and found it clear to follow. He also said that it 'Helped to provide point of discussion and made tutors aware of students' interests and skills' and it 'Helped put student at ease'. He also confirmed that a confidential reference had been supplied by the school and that this included additional essential information about pupils' medical condition, their attitudes to course work and an assessment of their potential 'ability to cope with extraneous demands of attending college e.g. independent travel'.

We have little evidence that potential employers etc, asked to see summative documents before interview or that they were used in selecting candidates for interview. Pupils and school staff were of the opinion that such documents were most likely to be useful in helping employers, in the context of the interview, to choose between candidates with broadly similar qualifications on the basis of other kinds of information such as personal interests or qualities. The little evidence we have supports this view and endorses the suggestion that the pupil's statement is one of the most important elements.

(e) Timing

During the pilot phase many of our case study schools realised that one reason why summative records were so infrequently used had to do with the timing of production. In so far as applications for further education places, and some jobs, had to be made early in the fifth year, summative records issued in the summer term, or even in the second half of the spring term, were of little use. The general trend therefore was to bring the process of statement writing forward so that at least some elements of the document could be made available earlier. We consider this move to be entirely appropriate but recognise that it has implications for the organisation of schools. For instance, if tutors are not to be put in the difficult position of writing and discussing statements with pupils they hardly know, it would seem important that tutors should continue with the same groups of pupils throughout the fourth and fifth year, at the very least.

1.3.5.
Use of computers

As already indicated, some of our case study institutions considered the use of computers at formative and summative stages in recording processes. This was conceived principally as a way of developing cost-effective and cost-efficient RoA systems for whole school use. In particular it was thought that computers could save time and money in the preparation of printed summary documents, although the potential to facilitate pupil-teacher discussion, by enabling quick access to up-to-date assessment information, was also recognised.

61

Records of Achievement

A major distinction between patterns of software design and use related to the nature of the RoA system developed, and in particular the distinction drawn earlier between **comment bank** and **open-ended or free prose** approaches to statement writing. Most of the software packages developed were tied to comment banks and all the criticisms levelled at this approach, and mentioned at various points throughout this report, can be construed as evidence of the limitations of computers. However, in this connection we feel that it is important to point out that problems arose, not out of the use of computers *per se* but out of the limitations of the systems they were expected to serve.

Computerisation of the comment banks was, from the beginning of the RoA project at Daisy High School seen as the most promising means of storing and retrieving records of assessments. The member of staff in charge of computer-aided administration carried out a review of available micro-computer software for this purpose, and concluded that nothing was commercially available which would suit the school's requirements.

Accordingly, this member of staff himself took on the task of developing a computer programme for handling the system's comment banks. The overall objective was to enable the speedy and automatic production of summative documents, and the programme, which was developed in-house, was used for the production of summative RoAs for fifth year pupils in 1987. Staff administering the comment banks ticked the comments applying to a particular pupil and these were then stored as coded data by the programmer. These codes in turn called up the text of the selected comments, which were then printed out department-by-department on each student's RoA.

It must be said that a number of serious difficulties were met during the course of computerisation. First, the member of staff carrying out the task of programming and data storage and retrieval found that these tasks involved dozens of hours work. The promise of a time-saving method for handling the comment banks was therefore not realised, at least in the short term. Second, the member of staff received no outside support for his efforts in terms of expert programming advice, which resulted in a great deal of time being taken up with experimentation. Third, the programme which was eventually devised was found by many members of staff to be unacceptably inflexible. For example, it initially imposed constraints on the number of characters (letters) permitted for each comment, and on the number of comments permitted to each department. These constraints were subsequently relaxed slightly, when it was discovered that departments could be allowed more than one line of text per comment, and more than one bank of 20 comments. These modifications notwithstanding, the sense remained among some members of staff that the computer programme was essentially constraining rather than enabling. In the third year of the project this programme was replaced with one developed in a neighbouring LEA, and the little evidence available relating to this suggests that many of the problems associated with the original programme had been resolved.

At Begonia School the cost of professionally producing a set of RoA documents for one cohort of pupils proved to be very expensive (to produce 500 sheets per subject cost £500). This was regarded as prohibitive. There were funds available from outside the school for the production of RoA documents and so computerisation was seen as a more economical means of production.

It was envisaged that subject-specific RoA documents would be replaced by computer-produced documents. This process began in the 1986–87 year. Early problems encountered were with the RML Nimbus system (four computers). The school began by attempting to use software used as part of the CISSA (Computers in School Administration) project for RoA purposes. This dual usage of the software created a number of problems. One problem was that the computer database could not be made to work. When a print-out of a pupil's file was requested, it was shredded by the computer. There was also a directive from the treasurer of the local council that teachers should not have access to the CISSA system because of its dual usage. Problems with software led to three departments developing their own programmes although these were all for different computer systems. The plan was to put all academic RoA schemes onto the RML Nimbus system and in December 1987 this had almost been achieved. Those who had developed computer programmes helped other staff to develop descriptors suitable for the computerisation of RoAs and whilst staff were obviously grateful for the help offered by their colleagues there was some criticism that the school did not receive any expert help from outside the school, the feeling being that it was unacceptable to expect other teaching staff to be on hand to offer advice and sort out problems. The local RoA project employed an information technology officer but her role was described as that of a facilitator — that is, she took staff to see pupil record programmes in operation — rather than as someone who could offer practical help at the level needed by the school. The local RoA project reportedly promised suitable software throughout the project but this never arrived. At the time of writing Begonia School was investigating a pupil records package (SIMS) which EMRAP promised to fund should the school wish to pursue this option.

The in-putting of the banks of descriptors developed by departments at Begonia was carried out by departmental ancillaries if available and by office staff who were offered overtime paid for by the school.

In order to facilitate the entering of assessments on the computer when interim statements were produced, a timetable of access was devised. This did not appear to cause any major problems, although there was some contention over who should type in such information. It was reported that the head teacher had said that staff should type in assessment information since this was equivalent to writing reports. However, there was some resentment from staff at being asked to do what they perceived to be clerical work.

With the computerisation of academic RoAs at Begonia came the decision that free-written teacher comments should not appear on computer-produced RoA statements. The argument was that the descriptors should say all that staff wanted to say and that if the option of writing a free comment was made available the descriptors would not be improved. However, a number of heads of department interviewed felt that their staff would like to make a free-written comment since this would make the RoA more personal and individualised. The PE department were particularly concerned about this decision. Early versions of the PE department's RoA schemes for third and fourth year pupils included sections for club involvement, county team representation, details of colours and awards received. Details of such activities were free-written by staff. However, following the computerisation of academic RoA schemes and the accompanying use of comment banks, a decision was taken at school level to exclude free-written comments on these computerised documents. As the development of comment banks covering the range of pupils' sporting, out-of-school or extra-curricular activities was felt to be untenable, such information did not appear on documents issued, much to the chagrin of the department.

It is conceivable that a comment bank system could be developed which overcame most of the problems associated with an essentially mechanistic approach to the portrayal of the whole child. If this were the case, and with expert support, computers could undoubtedly be put to good use in the service of comment banks. For the time being however the best evidence we have of successful use of computers is in association with software packages designed for use in more open-ended forms of recording.

The use of computers at Nasturtium Tertiary College became an increasingly important part of the RoA project in the college. Towards the end of the pilot phase there was every indication that all records would eventually be entered on BBC micro-computers for production of interim and summative reports. Those students and tutors who had experimented with the programme developed in-house were fulsome in its praise and claimed that by being student-friendly it had a strong motivating effect.

The system that was developed is particularly interesting in that it did not depend on prescribed comment banks; instead statements were generated by students, in free prose, using headings, brief course component outlines or criteria as prompts. The computer programme required course tutors to set up student disks with headings of course areas and a brief description of course content. A simple, non-expert user guide was produced to enable them to do this and co-ordinators conducted induction sessions. When disks had been prepared — one for a group of four students who each had their own password — students then entered their achievements under these heads, updating information as required. There was a limit of 5 or 6 lines per entry so the computer overwrote previous text rather than adding on. This had the advantage of forcing students to summarise achievement as they went along rather than leaving this task to the end. The resulting self-reports were intended to become the focus for one-to-one discussion in personal and course tutorials where they would be negotiated with staff. Students could choose to change their records in the light of such discussions and in order to secure their tutor's signature. However, since they held the password to their files, ownership and control symbolically and substantively rested with them.

One element of the summative RoA issued to fifth year pupils at Wisteria School was a curriculum vitae written by pupils using a programme designed by two members of staff at Wisteria. The programme was based upon the computer's word processing facility, was fairly short and simple to operate, and was devised to enable pupils to complete their CVs and print them out in a single tutorial lesson.

The distinction we have drawn between computer support for comment bank systems of recording, and computer support for open-ended systems of recording, is reflected in the principles underlying the software packages developed to support these systems. In essence, the comment bank approach used data-base type programmes whilst the open-ended approach used word-processing type programmes. We are inclined to conclude from the evidence currently available to us that the latter have the greater potential within the RoA initiative. However, we are aware that further investigation of experience elsewhere could easily persuade us otherwise. For example, some data-base software packages allow substantial entry of free text, and these could well be adapted for RoA purposes, as the following indicates.

Towards the end of the pilot phase at Magnolia Girls' Grammar School some considerable interest was generated in a computer programme developed by a colleague in another pilot school within the LEA. The local co-ordinator had been most impressed, believing that it could save time. He had therefore promoted its use in other schools. The programme was in two parts: one for academic recording; the other for personal recording. Both were based on a comment bank concept but, rather than comment banks being prescribed, schools were expected to devise their own, based, for instance, on departmental objectives. Levels of achievement were expected to be identified and staff were expected to select appropriate comment codes on grids. However, it was also possible to input numerical grades, individualised comments and pupils' self-assessments. In this way the software programme combined comment bank and free prose approaches and enabled the production of an output statement in which idiosyncracies could be added by agreement.

There is clearly still much to be learnt in this area, and we therefore welcome the opportunity for further development and evaluation provided by the proposed extension of pilot work.

Substantive RoA themes and issues

In this section we examine the way in which schools have attempted to integrate their RoA systems with other assessment, recording and reporting procedures, both within the same school and in their liaison with other schools.

A number of ways have been explored by which RoAs can facilitate schools' relationships with one another. The 1984 Statement of Policy implied that, in a fully operational national system, pupils who change schools as a result of moving from one part of the country to another should be able to take their accumulating RoAs with them and resume the process of recording at their new school. In this way RoAs could potentially enhance the transfer of records that is already an established element of such a move. Unfortunately we have no evidence relating to this aspect of between-school links from our case study work. The little evidence we have in this area relates to the way in which schools have attempted to build RoAs into more conventional transfer arrangements, either **from** feeder primary or middle schools if they are secondary or upper schools, or **to** upper schools if they are middle schools.

There is clear overlap between the liaison aspects of pilot schools' RoA-related work with their feeder schools and the work which some schools have carried out in helping to develop a primary school RoA which we discussed above in 1.3.1. Indeed those (relatively few) case study schools which have actively sought to collaborate with local primary or middle schools to develop a primary or middle school RoA have done so to strengthen what they felt was an already close relationship between the secondary school and its feeder schools, and to improve the quality of information provided to the secondary school about in-coming pupils, as much as to ensure that younger pupils were given opportunities to gain experience of RoA-type recording processes before they entered secondary or upper school.

1.4
Co-ordination and continuity

1.4.1
Between-school links

During the year 1986/87, the head of lower school (years 1 to 3) at Daisy High School held several meetings with the headteachers of local junior schools for the purpose of devising a system for summarising the achievement of in-coming pupils. This effort was built upon an already strong relationship between Daisy and its local feeder junior schools, as well as an established network of junior school headteachers in the area. The part played by the head of lower school was crucial in selling the idea to the heads of local junior schools.

The group of junior school heads themselves devised the pro-forma which was eventually used for pupils transferring to Daisy in 1987. The pro-forma was a single sheet of A4, with tick boxes alongside a range of personal qualities and space to record activities in and out of school. A pro-forma was filled in by 4th year junior teachers for each pupil transferring to Daisy from the participating junior schools. Teachers were able to indicate the level of maturity reached in relation to each personal quality listed by entering two, one or no ticks. In the junior school where the headteacher was interviewed by the evaluator, teachers met each pupil briefly in order to collect information for the sections of the pro-forma on activities in and out of school.

At Hawthorn R.C. Upper School the head of the fifth year in 1986/87 became the head of year for the new intake in 1987/88. In preparation for this she visited pupils in the main feeder middle school and took each class on three or four occasions before they transferred to Hawthorn. She decided to use these opportunities to work on the production of positive profiles to be brought with them on entry to the upper school. She encouraged feedback from both parents and children; this gave the impression that the exercise had been successful. In her judgement pupils of this age (13), in contrast with her experience of those who were older, had little difficulty finding positive things to say about themselves and by doing so they created a collective image of themselves which gave staff high expectations.

65

1.4.2
Within-school links

Of critical importance have been the links between RoA and other forms of assessment and reporting within the school.

(a) Assessment practices and policy

On the assessment front, RoAs have been introduced at the same time as a number of other initiatives — most notably GCSE, but also e.g. TVEI, LAPP and new subject-based curricular programmes, all of which have in many schools stimulated a fundamental re-appraisal of the principles underpinning routine assessment practices.

This diversity of new influences on teachers' assessment thinking and practice raises two sets of issues. First, schools have felt a growing need for whole-school assessment policies which attempt to integrate and reconcile the assessment principles informing these new initiatives. Schools are coming increasingly to the conclusion that RoAs hold the potential to provide this urgently needed integrating function, in their emphasis upon e.g. pupil participation in assessment processes, positive portrayal, widening notions of achievement, non-competitive indices of attainment, description rather than measurement, and explicit objectives and criteria.

Few of our case study schools attempted to formulate a whole-school assessment policy, and those that did discovered that the course was not smooth. On the one hand, where such a policy was devised by a few members of staff working in comparative isolation, and from a starting point of (laudable) principles rather than existing good practice, the new policy had little real impact.

On the other hand, where thorough consultation took place, and a genuine attempt was made to ground the whole-school assessment policy in existing practices and consensus, the way was hampered by widely differing levels of expertise and confidence, classroom practices, and conceptions of what a whole-school assessment policy ought to be and do.

In the autumn of 1986 an Assessment Working Party was established at Wisteria School. The aim of the working party was to discuss the feasibility of a whole-school policy on assessment and, if found to be feasible, to construct such a policy. Following the group's start in December 1986, it explored a number of different avenues, but several factors conspired to prevent the group from arriving at a promising basis for policy. Pressure of other work continually intervened, making it impossible to establish a sustained pattern of work for the group. The group also experienced difficulty in agreeing on e.g. terms of reference and definitions of terms, and members of the group felt a certain amount of diffidence about their ability to handle technical assessment concepts. Moreover, a tension emerged early on between, on the one hand, the wish of the working party to consult departments and to act as a relatively neutral co-ordinating team for an emerging consensus, and on the other hand the preference expressed by most heads of departments for the group to prepare a draft policy upon which they could then comment.

Essentially the group found it difficult to envisage what a whole-school assessment policy would look like. Should it be a set of guiding principles, such as an embargo on normative grades and a privileging of pupil self-assessment, or an agreed set of cross-curricular skills and concepts which all departments should foster and record, or simply a common format? In the summer of 1987 the working party referred its work back to the Senior Management Team, which at the time of writing had yet to make substantial progress in developing a whole-school policy.

The other assessment issue raised by the diversity of new initiatives among which RoAs find themselves is the more practical one of the time implications of many new systems being introduced simultaneously. Clearly, as long as RoA-related assessment procedures and other assessment procedures (especially for GCSE) are perceived as distinct assessment exercises rather than parts of an integrated whole, they will place an unreasonable burden on teachers. A complaint we often heard in case study schools was that teachers were spending so much time carrying out assessments that they had little time actually to teach. A whole-school assessment policy might go some way towards rationalising assessment demands, and the common principles underpinning various new initiatives will assist in this process. However, such a policy will only be a start. Our evidence suggests that, given supportive opportunities to think through the practical correspondences between the various assessment demands they face, particularly relating to the academic curriculum, teachers can devise effective systems for integration. On the one hand, formative recording processes in academic subjects can grow naturally from assessment criteria and continuous assessment procedures for GCSE.

On the other hand, coursework profiles devised under the aegis of the RoA project can be used to satisfy continuous assessment requirements for GCSE. However the problems should not be under-estimated.

Evidence from the four G Component subjects at Periwinkle School, OCEA, reflected a variation in the degree to which links have been made between Records of Achievement and GCSE. English found a ready match between OCEA criteria and demands of GCSE and found the two developments mutually supportive. Maths were 'battling with the plethora of objectives for assessment in years 1 to 3' and had, as yet, not addressed how they might be brought together in years 4 and 5. In science, where the OCEA criteria had been introduced for the first year, there seemed to be considerable disquiet as to whether it would be possible to cope even with the OCEA criteria let alone marry them to the GCSE, which are already extremely demanding — 34 criteria for science, for example. In modern languages the position was still more extreme where the assessment criteria were considered to be too complex and too unwieldy for practical use.

The fact that each new year of secondees began to rewrite the criteria compounded the problem. Overall it appeared that to the extent that departments could map the OCEA criteria on to GCSE demands, the former would survive. Indeed G disappeared as such and became redesigned in terms of criteria for each subject including GCSE criteria.

And of course there is no reason why pupils should not receive credit of an academic kind for some of the more tangible outcomes of formative recording processes on the personal side. Diary-entries, for example, have been submitted in English Language coursework folders for GCSE, and pupils who prepared their personal statements using word processors have incorporated that work into projects for information technology.

Finally, it should be borne in mind that a considerable amount of rationalising of assessment work will eventually be achieved through the framework for national curriculum assessments. It remains to be seen whether subject teachers will feel that so many aspects of achievement would go unrecorded in that national system that additional assessment procedures need to be adopted, and whether those further systems can be integrated with the national one.

(b) Reporting systems

We have observed at several points in this report that there has been a gradual move within case study schools away from purely formative recording processes towards procedures which incorporate some form of summarising and public reporting function. Whether this has been a cause of the growing use of formative records for the purpose of reporting to parents, or a result of the great need in schools to rationalise time spent carrying out recording and reporting, the fact remains that in many of the schools we have studied, traditional reporting methods have gradually been replaced by the practice of sending home the products of various formative RoA processes, e.g. interim personal statements, coursework profiles or units of work and so on. A fuller description of this development and of some of the issues raised by it is given in section 1.3.3 (**Interim records**).

1.4.3.

Post-16 provision

The DES policy statement emphasised that the establishment of RoAs for secondary schools was the first priority but recognised that FE institutions or tertiary colleges might also find RoAs valuable in that some courses being developed for 17 to 19 year olds included profiles. Our evidence suggests that issues of continuity and co-ordination between RoAs systems, both pre-16 and post-16, became an increasing focus for attention as pilot work progressed.

As was the case pre-16, teachers who had contact with a number of new assessment and profiling initiatives post-16 began to recognise potential links in terms of recording principles, procedures or products. Many came to hold the view that the various systems should be brought into some coherent relation to one another, preferably within some common overarching framework, in order to promote consistency of practice and to avoid the confusion which a multiplicity of different records creates for pupils, parents and other users.

In case study institutions where this issue was considered, a common view was that the RoA initiative could, and should, provide the framework — **the umbrella** — for the rationalisation of recording systems. It was felt that RoAs were especially suited to fulfil this role because they either incorporated most of the principal characteristics of other systems, or they were moving in that direction. For example, the idea that records of achievement should be based on criterion-referenced assessments, associated with explicit learning objectives in modules or units of work, was compatible with developments in BTEC and CPVE profiling. There were other developments however which were taking records of achievement **beyond** other profiling systems, particularly in the area of student self-assessment and recording, tutor-student negotiation of records, and descriptive accounts of achievement supported by concrete evidence. It was felt therefore that other profiling systems could be subsumed under the RoA initiative but that much might be lost if the profiling components of CPVE and BTEC etc were simply regarded as equivalent to RoAs. Nevertheless, as the illustration below illustrates, it was not always felt necessary to retain the RoA label as long as the principles were internalised in practice.

The experience of Nasturtium Tertiary College was highly relevant to this theme because, of course, all main provision was post-16. Whilst it proved extremely difficult and time consuming to convince colleagues, the college co-ordinators believed that Records of Achievement had to become the umbrella which would incorporate many other forms of assessment and recording into a coherent system based on a single set of principles. Unless it did this it risked becoming just another initiative contesting for limited staff time. Interestingly, the co-ordinators thought it might be important to drop the Records of Achievement label, which indicated a separate and special initiative, as long as the principles and processes were internalised.

In essence the strategy they adopted was to reinforce the idea that course criteria (GCSE, CPVE, CGLI, BTEC etc) were appropriate frameworks for assessment but that they overlaid the process of assessmment and recording with RoA-inspired principles and procedures. Thus staff were both encouraged to give students more responsibility for their own assessment and recording and to engage in dialogue with students to agree assessments on learning objectives and identify areas and strategies for future improvement. Some of these processes were already familiar to staff through other initiatives although BTEC, CPVE and GCSE relied heavily on staff-based assessment and CPVE required open-ended student accounts to be translated into 'can do' statements for processing by the Joint Board. Both these things posed problems but, with sufficient will, the co-ordinators saw no reason why the continuous assessments required of such courses should not be adapted to RoA principles. Indeed they negotiated with the CPVE moderator a form of recording, including personal recording, in line with RoA practice in the college. It transpired that the Joint Board was quite willing to accredit local variation. An encounter with a moderator for a BTEC National Diploma, who could see nothing wrong with percentages if they gave him a quick way to discriminate students ('It saves all that writing and reading'), gave the impression that he might be harder to convince, although he appeared to shift his ground after being involved in an active INSET session.

Perhaps the greatest difficulty was encountered in relation to GCE A Level courses (and O level courses till 1987) which had no conception of units of work based on specified learning objectives which might act as criteria for continuous course assessment. Although the college co-ordinators spent a very great deal of time working with GCE tutors and lecturers, progress was slow and the best that was achieved was a course progress report, which though designed to be discussed and signed by both student and tutor, included no statements with evidence but simply letter grades relating to termly attendance, effort and achievement.

Whilst our evidence supports the idea that records of achievement provide an appropriate framework for establishing coherence and continuity among post-16 profiling initiatives, we would not wish to underestimate the potential difficulties inherent in the task of unifying profiling systems. As this illustration makes clear some fundamental tensions are likely to remain concerning the place of marks or letter grades in records, concerning the use of 'can do' comment bank approaches to the recording of competencies, and concerning student self-assessment vis-à-vis assessment by teaching staff. Of course, it is possible to take a view of what constitutes an acceptable RoA system in which these tensions do not arise because, for instance, the inclusion of letter grades is permitted or, apart from a personal statement of interests and extra-curricular activities, all records are based on teacher assessment only. Our evidence suggests, however, that institutions which have tackled the post-16 issue would be reluctant to accept this solution particularly if young people have been exposed to RoA systems in their secondary education which have encouraged a greater measure of participation in and responsibility for the recording process. This, in their view, would be a retrograde step in that it would deny rather than enhance the maturity of older students.

1.5
Pupil differences

Many of the questions which concerned us in our case study work related to the ways in which differences between pupils might affect their experience of RoA processes. Like many others working in this area who are committed to improving the educational opportunities of girls and of pupils from social and ethnic groups who have generally been less successful within the education system, we have been interested to examine the extent to which participation in RoA systems might serve to improve such opportunities. We have also had an interest in the ways in which pupils of different ages and different levels of attainment might vary in their ability to participate in RoA processes, and the extent to which those processes have been adapted to meet different needs.

Like a number of other issues in the area of records of achievement, most of these questions relating to pupil differences were long-term, research-type questions which were difficult to resolve within the terms of reference framing our evaluation. It has, for example, been difficult to isolate one dimension of pupil difference (such as gender) from others (such as writing ability), and to isolate RoA processes as a causal factor. A great deal more research needs to be done before we have satisfactory answers to these kinds of questions, but we hope that our work may at least begin to cast some light upon them.

1.5.1

Age Because of the diversity of starting points for RoA projects in schools, we have been able to observe the experience of pupils spanning the whole secondary age range. One difficulty we face in analysing our evidence, however, is that the period of our investigation has not spanned the full five years of secondary education for any one group of pupils. We have certainly found, not surprisingly, that many pupils who have been introduced to the principles and practices of recording in their fifth year have found it difficult to understand the programme and to support it enthusiastically. However, this is a management issue concerning the phasing in of RoA systems, and one which we discuss below (see 2.1.1.).

It remains to be seen whether, as so many teachers we have interviewed believe, pupils introduced to these principles and practices early in their school careers are able by the time they leave school to participate in the processes with confidence, understanding and skill. We have found that, despite the difficulties which younger pupils have experienced in recognising and reflecting upon their achievement, younger pupils have taken to the processes of recording with more enthusiasm and less self-consciousness than older pupils. However, it does not necessarily follow from this that these younger pupils will be less inhibited, as fourth and fifth year pupils, than pupils who are not introduced to RoA processes until the fourth and fifth years. The culture of self-deprecation which seems so prevalent among teenagers may well be more powerful than the impact of participation in RoA processes at an earlier age. More knowledge in this area will be available as more and more cohorts of pupils experience a full five years of RoA-related work.

1.5.2

Gender Our evidence relating to girls' and boys' experience of RoA processes — and in particular teacher-pupil discussions, statement-writing and diary-keeping — is unexpectedly consistent. Bearing in mind that differences are rarely very sizeable, and that generalisations are difficult to draw, we have found that these RoA processes tend to appeal to girls more than to boys. In teacher-pupil discussions they are generally more forthcoming, better able to appreciate the purpose of the exercise, and more skilled at sharing responsibility for the conduct of the discussion. In this sense teacher-pupil discussions may capitalise upon the particular strengths of girls in the secondary age range such as social awareness, maturity and competence in one-to-one conversations.

An interesting relationship emerged between gender and pupils' preference for group or one-to-one tutorials in the analysis of evidence at Ysgol Eithin, Wales. Whereas boys and girls gave good ratings to both group and one-to-one, many boys — but no girls — rated the groups higher than the one-to-one tutorials:

Preference	% Girls	% Boys
Rating tutor groups and one-to-one equally	46	14
Rating one-to-one higher than groups	54	43
Rating groups higher than one-to-one	0	43

In interview with the evaluator, the fourth year form tutor at Daisy High School, whose class was studied as part of the close focus, shared some observations she had made about boy-girl differences during tutor-pupil discussions of activities in and out of school. With some pupils — and according to this tutor these pupils tended typically to be girls — there was if anything more material to discuss than could be accommodated in the short time available, and the pupils took a share of the responsibility for keeping the interview flowing. With other pupils — and, inversely, this tutor found that boys much more often fell into this category — generating relevant information for the folder was 'like getting blood from a stone'. The tutor accounted for this in part by boys' comparative uninvolvement in formal organised activity, but also by the tendency on the part of some boys apparently to be virtually oblivious to the whole point of the exercise, to find one-to-one discussion with a teacher almost unbearable, especially with the topic of discussion being themselves, and to treat the interview as an ordeal to be endured rather than as an opportunity to chat.

In terms of statement-writing and diary-keeping, girls tended generally to demonstrate a clearer understanding of the purpose of the exercise, to write more and in greater detail than boys, and to place more value on both the process and the product. Here again, it may be that these kinds of activities build on the more positive orientation of girls towards writing, their enthusiasm for comparatively private and solitary activities, and their apparently boundless capacity to plan for nostalgia.

Staff at Buttercup R.C. High School reported that girls were more diligent than boys when it came to continuous recording in logbooks. One commented that the girls wrote in minute detail about what they had done, whereas some boys had not bothered with them at all. This could create problems for tutors, one of whom reported that there was nothing for pupils to refer to in reviews when the progress profile had to be completed. This tutor asked pupils to complete their progress profile before the review so that there was something to discuss. Tutors also reported that girls used the personal reflection section (green pages) of the logbook more often than boys, and were more ready to commit their thoughts to paper. This was borne out by evidence from pupils interviewed. Only one girl of six interviewed reported that she did not use this section of the logbook, although another reported that she only used this section as an 'appointments diary', being under the impression that staff had access. Of seven boys interviewed only one said that he used the green section. Some of the boys interviewed were concerned about personal information from this section being read out in class by a pupil who managed to get hold of another's logbook. Two of the boys could not see the point of recording information that no-one was going to see: one commenting that he could see no value in recording personal thoughts and reflections. He said that he would use the section to record a particular exciting event. These two pupils did not attribute any intrinsic value to the recording of personal reflections and as no-one would see and take note of what was written there was no extrinsic value either.

In this sense we can speculate tentatively that RoA-processes, and in particular teacher-pupil discussions, statement-writing and diary-keeping, may go some way towards rectifying the imbalances between boys and girls which have been associated with such traditional school processes as large-group and whole-class question-and-answer sessions and concentration upon propositional knowledge as opposed to personal experience.

On the other hand the growing tendency to use formative records to serve public reporting purposes might erode this possible advantage, since boys tend to have a keener sense of external audience in their writing and to concentrate on concrete achievements rather than feelings and relationships.

In Campion Upper School, Suffolk, the school's own evaluation of the RPA with the fifth year in 1986 revealed no significant difference in overall response between the sexes. However, in the evaluator's interviews with a small sample of pupils in the fourth year in 1986/87 there was some indication that boys had a more instrumental view of RPA. Value for boys appeared to attach primarily to its usefulness in securing employment whereas girls ascribed principle value to its formative potential i.e. helping teachers to know them better and helping them to know themselves. Thus girls appreciated the opportunity to talk one-to-one with their tutor whilst boys saw more value in the summative statement. This divergence of view was also reflected in pupils' assessments of the Guidance Programme in which RPA work was located: girls found it helpful but boys interviewed thought that much of it was a waste of time. Added together, and not unexpectedly, these observations suggest that girls attach greater value to the pastoral (or affective) curriculum than do boys.

Two final points relating to RoAs and gender are worth noting. First, there is the question of whether it is preferable for pupils to be interviewed by reviewing tutors who are of the same sex.

In the teacher-pupil discussions observed at Campion Upper School there was little indication of difference according to whether the interaction involved boys or girls. However, one girl said: 'I don't find it easy to talk to teachers, specially male teachers — female teachers a little bit better — but there's nothing wrong with the system, I think that's quite good'.

There was little evidence at Buttercup R.C. High School to suggest that the sex of the reviewing tutor affected the interaction, although there remains the issue of the prudence of male tutors talking to female pupils in such a situation. One female pupil commented that although she found it easy to talk to her male reviewing tutor she would have preferred an interview with a woman. However, other evidence suggests that the reaction of male and female pupils in reviews was independent of the sex of the reviewer. Both a male and female tutor reported that boys seemed to find it less easy to talk in reviews than girls, the latter being more forthcoming.

The issue here may be particularly acute in situations where male tutors are expected to conduct one-to-one discussions with Asian girls. Alternative arrangements may, however, be difficult to make within the existing tutorial arrangements of most schools.

The second point relates to the common assumption that girls tend to underestimate achievement in self-assessment, whilst boys overestimate achievement. We have some evidence to support this.

It was reported by home economics staff at Jasmine High School that when making self-assessments girls underestimated their achievements whereas boys overestimated. This was felt to be because girls felt they should be good at the subject. A member of staff from another department reported that high attaining girls frequently underestimated their achievements. This teacher also reported that girls tended to go along with the teacher's assessment because the teacher was regarded as 'all-knowing'. Boys tended to put their point of view more strongly.
A male teacher reported that he found girls more honest when making self-assessments, but boys seemed to respond better in reviews, in that they talked more than girls who seemed rather quiet.

However, most of our evidence led us to conclude that the undoubted tendency to underestimate achievement was shared roughly equally by both boys and girls and indicated, not a gender-specific phenomenon, but a cultural norm associated perhaps with acquired national characteristics such as 'British reserve'.

We have to report that our data in this area were insufficient to permit even speculative remarks of a general nature on the experience of pupils from ethnic minority backgrounds. This is due in large measure to the way in which the sample of schools and schemes taking part in the pilot programme was skewed significantly away from inner-city areas with large proportions of ethnic minority pupils (see Appendix 2). Added to this have been the constraints, of both resources and methodology, on our ability to pursue research questions in this area. For the record, however, we present the little evidence we have in the following extracts from our case study reports.

Sunflower High School had a high proportion of ethnic minority pupils. Staff reported a number of issues relating to Asian pupils. One tutor reported that some Asian pupils found interviews with staff very demanding linguistically. Although there is always the issue of the prudence of girls talking to male teachers, there are added cultural problems for Asian girls talking to male staff and possibly for male pupils talking to female staff. There may also be problems when talking about out-of-school activities, since so much of some Asian pupils' life outside of school revolves around the Mosque.

Of the nine Student Statements from summative London Record of Achievement portfolios which were available for analysis at Hydrangea School, ILEA, the one which was longest and most detailed was by a student who was Asian and female. Indeed this pupil's portfolio was in every sense the most complete of the nine, with more samples of work, more formative Unit Credits, and more evidence that the exercise had been understood and supported. However, this evidence is much too limited to enable general conclusions about issues of ethnicity.

Numbers of pupils from ethnic minority groups at Hawthorn R.C. Upper School were very small although the school attracted some pupils from families from Catholic or Orthodox national cultures e.g. Italians, Greeks. Some of these families, together with a few from Asian and Oriental ethnic groups, were involved in the restaurant trade. Two such individuals were among those in the group for close focus study and both indicated that their opportunities for developing, and therefore recording, out-of-school interests and activities were limited by their involvement with the family business. They were also both highly motivated towards academic success and said that their own time was occupied almost exclusively by restaurant work or homework.

One boy at Cornflower Special School came from an Asian family for whom English was a second language. The problems he experienced in recording were immense. Communication with staff was often difficult and he did not demonstrate any real understanding of purpose. Continuity was also difficult to maintain: his medical treatment consumed much time in school and for a long period during the course of this study the whole family went missing and could not be traced. Personal recording was very dependent on the support of the OCEA assistant who took great pains to elicit relevant information but who did not always succeed. The interim summaries produced represented a genuine attempt to encourage this boy to speak for himself, with the OCEA assistant acting as mediator, although they could not be said to conform to any conventional definition of a record of achievement. The language dimension therefore created some considerable difficulty although in this case the communication problem was not simply a problem stemming from ethnicity; the nature of the pupil's illnesses were a very significant contributory factor.

Other pupils of ethnic origin in the OCEA target group appeared not to have such great problems in communication although cultural norms, especially in relation to Asian girls, appeared to limit the range of experiences and achievements on which they were able to draw. For instance, it was noticeable that the out-of-school experiences of one Asian girl all revolved around the home and the Mosque. In this light, the familiar expectation that the personal record should include sports, membership of organisations etc began to look peculiarly ethnocentric. However, here again the severity of the pupil's illnesses was undoubtedly a contributory factor since it limited what she was able to do both within school and outside.

In this connection it is of some interest that staff perceived a relationship between ethnicity and the severity of pupils' medical condition, which in turn affected their experience of recording achievement. Although Cornflower Special School served a city which has a large ethnic community, mainly Asian, calculated as more than 30 per cent of the total population, this proportion was not reflected in the population of the school. From 1985 to 1987, less than 15 per cent of pupils were from ethnic minority groups. One member of staff articulated a theory that some Asian children were kept within the family and escaped being 'statemented' until the situation was so serious that the family was no longer able to cope alone. Thus ethnic minority children in the school were more badly handicapped by their illnesses than many of the rest. Significantly perhaps, all three children who died in 1986 came from the Asian community.

(b) Welsh language dimension

Welsh was the main language of the catchment area of one of our case study schools. The insights gained with respect to the language issue in this school are summarised in the following extract from the case study report for that school.

The evaluation of the pupil perspective at Ysgol Eithin exposed an important issue concerning pupil access to their choice of language. This issue presents a problem for the achievement of optimum success for profiling in bilingual communites. At Eithin, within the very favourable overall pupil view of their tutorial arrangements, one criticism concerned the medium rather than the content of the transactions. We found that a significant number of pupils wished to communicate in their preferred language and not the language imposed by the tutor. Welsh is the first language of the majority of the community served by Ysgol Eithin. Though an increasing number of English-only parents have and are moving into the area it would be virtually impossible for their school-age children not to possess a coping ability in Welsh after a relatively short period in the locality. When we interviewed these 'immigrants' they were quick to make it clear, usually by answering in Welsh to an initial greeting to them in the language, that they must conduct the interview itself in English. Apart from our giving them a choice of language medium for the interview, we did not ask the thirty-four pupils we met any question on the topic of language in relation to their records of achievement experience. When we analysed our data, however, pupils' written comments in response to the self-completion exercises we had set them, and our own notes, showed it to be an issue. We found that 18 per cent of our interviewees had made a written or oral statement about the language used in their tutorial transactions.

All but one of the pupil comments were concerned about having to speak English when Welsh would have been preferred. The same issue applies, however, whether Welsh or English is the 'enforced' language, if given the opportunity the pupil would choose the alternative language. The essence of the student 'complaint' contained in these comments is that they feel disadvantaged when they are not able to speak in the language they feel most comfortable with, and which can communicate what they want to say, learn and feel.

There is one important sense in which this problem is specifically a Welsh language resource issue. Whereas all tutors can speak English, not all can speak Welsh (the indigenous language of the area!), hence the appearance in our data of a significant pupil perspective on language. Though only 15 per cent of our interviewees 'complained' about lack of access to Welsh, this in fact represents a very large proportion of the minority of pupils in the year group who had had their tutorials conducted by a monoglot English tutor. The more general resource issue is that whereas everywhere in Wales all tutors can meet pupil choice for English as a medium for tutorial transaction, not all pupils' preference to have tutorials in Welsh can be met even in areas where Welsh is the main language of local society.

Oral communication is central to active tutorial work whether on a group or individual basis. The language medium in which pupils feel most skilled and confident is therefore a crucial element governing the learning outcomes intended by such work. If pupils are disadvantaged by the language used they cannot play their full role in the development of their record of achievement. Most importantly , where there is language mismatch between tutor and pupil there is unlikely to be in the crucial one-to-one tutorials a growth of the required degree of rapport with the trust and candid interchange necessary for the development of a realistic assessment of abilities and pupil self-image. In short, the mismatching of language medium to pupil need could inhibit the full accuracy and scope of the record of achievement.

We see three implications arising from the 'language choice problem' for a school's management of records of achievement in the bilingual setting:

First, tutors must be linguistically competent to undertake the work, i.e. they must be bilingual. Where schools have a policy of tutors staying with the same group from years one to five, this will need review if appropriate staff resources are to be accurately matched with RoA activities in years four and five.

Second, pupils should have a choice of which language they wish to use in one-to-one tutorials, and adequate recognition must be made of all pupils' language primacies in tutorial work. The principle of pupil contribution to, and ownership of, the record of achievement is unfulfilled if pupils' felt language needs are unmet.

Third, tutors and pupils need proven Welsh medium materials to implement records of achievement optimally in the bilingual setting. In particular, these are needed for active tutorial work.

(c) Social circumstances

The main issue emerging from our evidence which relates specifically to RoAs and social circumstances, and not to the general problem of education and social disadvantage, is the apparently wide variation among pupils from different social groups in the kinds of out-of-school activities, experiences and interests which they may bring to the recording process.

The evaluation carried out by Campion Upper School, Suffolk, with the fifth year in 1986 revealed no significant differences in response to RPA between working class children and more middle class pupils drawn from some of the villages. However, it was a cause of considerable concern to staff that pupils living in remote rural areas and less affluent homes had less to write about in their profiles because fewer opportunities were available to them. They appeared disadvantaged both by the limited range of facilities available to them and by the time taken in travel which diminished the time available for other forms of participation.

Although the rhetoric of RoAs emphasises that positive recognition should be given to the way in which **all** pupils spend their time outside school, regardless of their social background, this ignores the powerful forces operating within schools which communicate to pupils the kinds of activities which are, and are not, acceptable. Notions of 'cultural deficit' are deeply ingrained, and teachers are quick to impose (unintentionally) those notions on pupils and to accept the reports by pupils from 'disadvantaged' or rural backgrounds that they do nothing out of school worthy of recording. On the other hand, teachers may be justifiably torn between their wish to draw out and give real recognition to their pupils' out-of-school achievements, and their concern that outside users of a RoA document may react negatively to descriptions of activities which attract social disapproval.

One fifth year tutor at Campion Upper School expressed the view that for three or four low-achievers, in her mixed ability group of 27, preparing a final statement of any value was almost impossible because there was very little that was positive that they could find to say. A slightly larger group, perhaps eight in all, had little or no concept of audience and could not understand how negatively employers etc. might interpret the things they insisted on recording e.g. 'I play darts for the Marquis of Granby'. In the case of one boy, he and she decided that the best they could do was to agree that he should not have a final RPA statement because such a document would either be misleading or too damaging.

The task of drawing out and placing genuine value upon the out-of-school activities of **all** pupils therefore involves far more than training in interview techniques, however important these are. An attitude of mind is needed which can accommodate the widest possible range of interests and experience, and an ability to communicate that attitude without condescension. Finding strategies to cope with the issue of user prejudice is even more difficult, and our evidence offers us no easy or immediate solutions. For this reason we can do little more than indicate this as an area for further research and deliberation.

1.5.4
Attainment
(a) Low-attaining pupils
(i) Problems in assessment

The most notable way in which low-attaining pupils might be disadvantaged by assessment frameworks devised as part of RoA projects has been in the tendency of the more structured assessment frameworks (e.g. comment banks and staged assessment systems) to be hierarchical in their treatment of attainment and therefore to prohibit some pupils from gaining many statements or stages, or indeed any at all. For example, some pupils are unable to reach the lowest stage of some staged assessment frameworks, and the small number of statements out of some comment bank systems which can be applied to low-attaining pupils contrasts dramatically with the large number which can be applied to other pupils.

Problems associated with assessment and recording at Cornflower Special School, OCEA, were considerable. In terms of assessments made by subject teachers there was an evident tension between wanting to give encouragement for effort and wanting to give an accurate assessment of achievement. The intentional or unintentional strategy the science teacher adopted to resolve this tension was to mark course work using marks out of ten and a general comment (e.g. good, very good), but to set and assess special assignments using OCEA criteria. Thus pupils had the satisfaction of having folders of work with high marks and encouraging comments, in recognition of the teacher's judgement that they were doing their best at the time, although according to his assessments on OCEA science criteria it was evident that they had difficulty in demonstrating some skills at even the lowest level. After making OCEA-type assessments on the basis of his observations of pupils' work, the teacher said that he discussed and agreed these with pupils. The contrast between routine marks and OCEA assessments appeared to go unchallenged by pupils, probably because they were expressed in such different ways. However, there was some evidence that pupils took the former as the true indicator of their achievement. This would explain why one boy confidently declared that he did not think his teacher thought he needed to improve, although this was patently not the case.
The science teacher was not unaware of this difficulty and had discussions within his science reference group about the possibility of adapting the OCEA assessment framework so that it encompassed lower levels of achievement and avoided any sense of failure that would be engendered in pupils who could not

demonstrate skills at the current lowest level.

Related to this, two further problems were encountered at Cornflower School. First, the nature of pupils' handicaps often made it difficult or impossible for them to manipulate equipment in the way necessary to demonstrate skill. Secondly, pupils with progressive illnesses were faced with the demotivating possibility that their level of performance would actually decline over time. Staff discussed with their colleagues in the county how these situations might be ameliorated. For instance, they thought it legitimate to accept pupils' instructions to others to manipulate equipment for them (even by means of nods and shakes of the head) as evidence of achievement. However, these issues were not fully resolved, especially the issue associated with progressively debilitating illnesses.

The head of special education at Daisy High School was very concerned about the implications of RoAs for pupils with special educational needs. First, she found it difficult to be both positive and accurate about some pupils in her department. More importantly, however, she felt that the actual number of statements a pupil received from the maths and English comment banks served as a normative reference point, since higher-attaining pupils received many more comments than low attaining pupils. She felt that learning for pupils with special needs was not progressive: they were able to demonstrate competence in relation to a skill one day, and be incapable of doing so the following day. This made it difficult to select comments from the bank with confidence. She also worried that RoAs might serve to disadvantage still further pupils with special education needs, and found it ironic that an initiative which had originally been conceived as a system of recording and certification for non-exam pupils should present such enormous difficulties for low attaining pupils. In the third year of the pilot a more sophisticated computer programme was adopted which enabled the length and number of statements to be considerably increased.

Whilst it is inevitable that assessments will differentiate between pupils, or at any rate enable differentiation between pupils, the implicit normative principles of these frameworks emphasises the differences between high- and low-attaining pupils and makes it difficult to realise the RoA purpose of giving positive recognition to all pupils when their records are compared.

One prominent feature of many of the RoA systems we have studied was the considerable writing demands they make on pupils. Diary-keeping, statement-writing, self-assessment and other methods of self-accounting, add to the already substantial requirements for pupils to produce written work at school. Moreover, the increasing use of portfolios which contain 'best' pieces of work extend even further this emphasis upon writing. We have already mentioned the possible ways in which this emphasis may give girls some advantage (see 1.5.2 above). Nevertheless, it is important to recognise that a large proportion of pupils experience writing difficulties of one kind or another — either through their relatively low level of language development, or problems with handwriting, or simply a dislike of writing — and that this will affect their attitudes towards RoA processes.

(ii) Writing difficulties

Several attempts have been made in case study schools to minimise the alienation of pupils with writing difficulties from RoAs. First, giving pupils the option of preparing their work on a word processor has in many cases broken through their antipathy (see also 1.3.5).

Second, schools have discovered ways of using pieces of writing prepared in the context of RoA processes for some other additional purpose such as GCSE coursework, which may in time rationalise the writing demands overall and provide greater integration of RoA processes with other school work.

77

Finally, teachers have enabled pupils with writing difficulties to dictate e.g. their statements, either directly to a teacher or into a tape recorder for later transcription. This has been transcribed to provide a first draft, thus overcoming the intial hurdle.

(b) High-attaining pupils

This section concentrates on the reaction of high-attaining pupils to the process of recording. We have evidence that this group of pupils seemed more likely than others to question the relevance of aspects of the process, in particular the personal element. Since high-attaining pupils tend generally to be prepared to conform to school expectations, they have on the whole participated conscientiously in RoA processes as they would in any school task which was expected of them. However, a number have questioned whether they would benefit directly from the process when they were single-minded in their intention of going on to higher education, and have in some cases resented the loss of revision time caused by participation.

It was the perception of staff at Hawthorn R.C. Upper School that the school had rather more high-attainers than low-attainers. The target group Head of Year offered two possible explanations for this: first, the Irish Catholic attitude towards education encouraged scholastic achievement; secondly, many aspiring parents appeared to choose church schools because they saw these as the nearest thing to independent schools. Given the expectations of parents, staff identified a problem in getting high-attaining pupils to attach any importance to a RPA emphasising personal achievement. Data collected in observation and interview supported this staff perception to only a limited degree. Admittedly, high-attaining pupils tended to go along with the RPA exercise because they regarded it as another required school activity. They would have preferred less time to be spent on it and they were uncertain of its relevance in terms of gaining access to higher education. However, they recognised the importance of personal qualities and admitted that they had gained from being made to reflect on themselves as whole persons. We encountered no resentment of the tasks, and the opportunity to talk on an individual basis with their tutor was valued, if only because it gave them a chance to ask questions about examinations. Acceptance of personal qualities as a legitimate dimension of recorded achievement seem to be related to the predominant ethos of the school. A definition of education was cultivated which clearly incorporated spiritual, moral and personal values — and pupils accepted this.

Any reservations on the part of high-attaining pupils may well become less strongly felt when RoA processes are extended into sixth forms and FE institutions, and when all pupils become better informed about possible users of RoAs other than post-school destinations.

It must also be said that some high-attaining pupils reported that they enjoyed the opportunity to talk to their teachers on a one-to-one basis about themselves. In this sense they placed particular value on the process as opposed to the product. This was especially true in the selective school in our sample, where opportunities for formal one-to-one discussion with teachers had previously been rare.

We have no evidence that girls at Magnolia Girls' Grammar School, who were mostly relatively high-attainers, considered the RoA exercise as irrelevant. Indeed some expressed an interest in having more time to talk with tutors about their personal achievements and qualities. They gave the impression that they expected to find benefit in something to which staff were prepared to commit time and effort.

We are now in a better position than we were in July 1987, when we published our interim evaluation report, to say something about the impact of developments in validation and accreditation on case study schools. However, our account remains mainly descriptive because the evidence is not yet available that would enable an analysis in terms of distinct models emerging **at school level**. Neither are we yet able to to draw general conclusions about the relative effectiveness of different approaches. Indeed, because most validation and accreditation procedures are developed at scheme level, readers are encouraged to refer to the more detailed analysis of accreditation schemes in Part Two of our report, which summarises reports from project directors.

We would wish to preface our remarks in this area by emphasising that there was no unambiguous definition of, or clear distinction between, validation and accreditation as concepts underpinning practice in case study schools. However, the term 'validation' appeared less frequently and we have no evidence of procedures designed **to validate the records of individual pupils**. The terms validation and accreditation were thus generally applied to the **processes** which schools have adopted to develop records of achievement. Nevertheless, when these processes involved examination of samples of documents, which were the product of the process, as it did in some cases, the procedure came close to **validation of records** in the stricter sense.

With regard to the range of experience of validation and accreditation at school level we encountered considerable diversity. In some schools the terms were rarely if ever used, reflecting the low priority of development in this area. In others, though no specific procedures had been developed, there seemed to be an assumption that 'quality control' would be ensured by a combination of fairly strict county guidelines regarding principles, processes and products, by in-service education and training, and by routine inspection by the county inspectorate. Elsewhere explicit and detailed procedures were initiated. Within this last group of schools, a common pattern, as far as one can be detected, was for schools to apply to the scheme for accreditation by preparing a submission based on a review of their current and projected practice and provision in the light of scheme criteria or principles. These submissions were then scrutinised by a specially appointed accreditation group composed of representatives from the scheme project team, the LEA's officers and advisers, and sometimes an examining board and colleagues from other schools. Often the procedure involved some members of this group visiting the school. If the school satisfied the group that it had met the criteria, or that it had established a development programme to achieve these ends, then the school was given accredited status which usually entitled it to use the crest or logo of the local authority or scheme. The following examples illustrate a number of variations on this theme.

Periwinkle School, OCEA, was identified as a pilot school for accrediting OCEA in the county. The notion of accreditation developed over the period of the pilot scheme. In 1986 the procedure was simply to register individual pupils from first to fifth year with the Delegacy on the basis of £12.50 per pupil registration fee and a fee for each subject, £2.50 for a G Component pack, which could be duplicated — fees payable by the LEA. No schools paid any fee during the pilot phase and the fee was subsequently modified to £4.05 which is the current rate. Periwinkle School decided to register pupils for maths in the first and second year, science in the first year, humanities in the first year, English in the third year and some fifth years. Since that time the OCEA scheme has itself been developed within the consortium and within the LEA to produce a more LEA- and school-based model of accreditation in which the Delegacy's role is perceived by the school to be progressively diminishing. The Delegacy's role will then be to provide external examination results for GCSE and for other curriculum development projects, graded assessments etc. as appropriate. Meanwhile, in the future, Periwinkle School is likely to be doing GCSE exams for several boards, and therefore it will be the schools responsibility to validate even the E Component entries on each pupil's summative statement. The procedure that has evolved in Oxfordshire involved a preliminary drafting of accreditation procedures by the OCEA accreditation officer and the local county co-ordinator. These draft procedures were discussed by both the OCEA secondees and the heads in the county. Building on the county commitment to periodic school self-evaluation, a procedure has evolved based on the nine OCEA accreditation principles which will be validated for each school through dialogue through the LEA officer responsible and the school. Each pilot school has an appointed 'friend' in the person of an advisor who will visit regularly to discuss the conduct of the Record of Achievement scheme in relation to the agreed accreditation principles. The school, in its turn, will produce a public statement in relation to these same principles. The Oxford Delegacy will keep a copy of the summative record of achievement on microfiche.

In January 1988 an OCEA accreditation team visited Periwinkle School to look at the county arrangements for accreditation. The team involved personnel from the Delegacy, from other LEAs involved in OCEA, the OCEA director and the OCEA accreditation officer. The visit was not intended to see whether a school should be accredited but to study the procedures that the LEA was evolving to provide for accreditation and to use this knowledge to develop more general insights about accreditation procedures. Some of the insights gained from this exercise are recorded in *Accreditation in Oxfordshire — Views from the schools*, which describes desirable and undesirable features of an accreditation process.

Desirable: Should be a system which accredits practice; should be a system that offers support in the interpretation and implementation of the accreditation principles; should be a system which ensures the accreditation principles are not undermined and that the credibility of OCEA is established and maintained.

Undesirable: Should not be a system that creates an alternative inspection system or unnecessary bureaucratic procedures; should not be time-consuming.

Thus it can be said that, at the time of writing, the accreditation procedures were still in the process of development. The principle of LEA responsibility and support — mainly the visits of a particular adviser plus the responsibility of a school to produce its own self-account – are the determining characteristics in Oxfordshire. There is a clear distinction between the accreditation of the institution and the validation of particular summary statements by the Delegacy. In this respect the school experienced problems with the Delegacy during the pilot phase in getting the necessary stationery and validating stamp in good time to issue summative statements to pupils. Our evidence suggests that whilst OCEA will provide the county umbrella for the initiatives, it will increasingly be perceived as an LEA, rather than a consortium-wide, scheme by schools with the Delegacy's direct role becoming increasingly marginal since it has no direct accrediting function.

Each school taking part in the Dorset RoA scheme, including Bluebell School, was expected to convene a Validation Board. Validation Boards were school-based groups, and members were formally appointed by the school's governing body. The functions of the Validation Board were (a) to assist the institution in the development of its recording system and (b) to assist the institution, the local education authority and the local examinations group in the process of awarding certification for the summative RoA.

The county guidelines (*The Role of Validation Boards in Dorset*) set out four stages of development which framed the role of the boards:

Stage 1: Declaring intentions: The teachers in the institution should declare publicly their intentions for (a) translating the aims and objectives of the county curriculum policy statement into practice and (b) translating the Dorset 'Principles for Recording Achievement' and the aims and objectives of the Assessment Policy Statement into practice. This declaration should focus on the curriculum, teaching methods and learning styles, assessment and recording, and administrative arrangements.

Stage 2: Granting approval: The Validation Board should examine with rigour the intentions declared by the institution in Stage 1 and should then either (a) give approval to the declared intentions or (b) indicate what should be done in order that such approval might be given.

Stage 3: Monitoring: At least once per year the institution should demonstrate to the Validation Board that the intentions declared in Stage 1 have been translated into reality. This monitoring should include some monitoring of classroom practice, conducted either by members of the Validation Board or by the LEA on the Board's behalf.

Stage 4: Certification: Once the previous stages have been completed to the satisfaction of the Validation Board the institution will be given permission to use the 'Dorset Record of Achievement', which will have the certificating crests of the LEA and the Southern Partnership for the Accreditation of Records of Achievement. The authenticity of the RoA will be underwritten by the Validation Board.
Members of the Validation Board at Bluebell included the chair of the governing body, the head teacher, the school RoA co-ordinator, the deputy head with responsibility for curriculum, a representative from the local examinations group, a member of the Dorset Assessment Team, and two representatives from employers in the locality who were also parents of pupils in the school. From time to time, other individuals attended meetings, including a representative of the local careers service.
Meetings of the Bluebell Validation Board commenced on a termly basis in the autumn of 1986. The Board acknowledged from the beginning that it would not attempt to grant

certification for the summative RoA issued to school leavers in the summer of 1987. It saw its task as a long-term one, and moreover as one principally concerned with educational processes rather than user-orientated products. The school for its part was prepared to issue an 'uncertified' summative RoA to school leavers (as distinct from one for which certification had been withheld) until such time as the Validation Board felt sufficiently well-informed and sufficiently satisfied with the school's internal processes to address itself to the validity of the summative RoA.
Most of the discussions which were held at the three Validation Board meetings which the evaluator attended in 1987 dealt with issues which roughly correspond to the first two stages of development outlined in the county guidelines quoted above. Although the Validation Board discussed a wide range of issues relating to the school's efforts to translate the county principles into practice, at no stage was 'approval' formally granted in the terms set out under Stage 2. Nevertheless, school staff continued to work on the assumption that they had the broad blessing of the Validation Board. Plans to begin a monitoring stage were discussed at the last meeting of 1987, but no evidence relating to this exercise is available.
One of the difficulties in describing the work of the Validation Board at Bluebell is that the board was expected, within the terms of the county guidelines, to define its own terms of reference, its own mode of working and its own procedures for validating the RoA-related work within the school. No evidence is available to suggest that the Validation Board at Bluebell ever formally agreed terms of reference for itself. As a result, discussions at board meetings tended to oscillate between formative process issues and summative document issues, and between the educational value of what the school was developing and its possible value to users outside school.

Jasmine High School, Wigan, was involved in NPRA Unit Accreditation as part of Wigan's involvement with the Partnership. The procedure for unit accreditation involved the design, by schools, of units of work in line with published NPRA criteria for validation. These units were pre-validated

by an LEA panel before being submitted to the Regional Validating Committee. If the unit was passed as validated it was taught by the school. Assessment took place and records and evidence of pupils' achievements were kept. When a pupil had successfully completed a unit, the school informed the NEA which sent an assessor to inspect the evidence. When the NEA was satisfied, it informed the LEA which then issued a statement of achievement. At the end of the whole course the NEA issued a pupil with a letter of credit giving the titles of successfully completed units. At Jasmine, a teacher was appointed as co-ordinator of the Unit Accreditation scheme.

Most departments in the school were involved in the unit accreditation scheme and a number of staff had elected to write their own units. The unit accreditation co-ordinator helped staff to write units and there was a strict format to be adhered to before being submitted for pre-validation.

At the time of writing the Clwyd RoA scheme operated by Honeysuckle School was accredited by Clwyd LEA but accreditation by the WJEC was under consideration. However, according to the LEA's RoA co-ordinator, the WJEC had said that they would not accredit schools involved in the Clwyd scheme (centre accreditation) until the scheme was brought into line with WJEC Criteria (based largely upon the 1984 DES Statement of Policy). The WJEC were only concerned about the inclusion of what it considered to be negative comments in the summative document. According to the LEA co-ordinator the WJEC had stated that if Clwyd agreed to withdraw these 'negative' comments from the final summative document the RoA would receive accreditation by the WJEC for the 1988 cohort of fifth year leavers. This was not acceptable to the Clwyd LEA, as it was felt that time would be needed to implement such changes since the comment banks were used formatively during years four and five as well as summatively, and since changes to the comment banks would have to be introduced with fourth year pupils beginning their involvement in the Clwyd scheme. The authority was also reluctant to make changes to the banks of comments prior to the issue of RANSC's final report. At the time of writing this matter had not been resolved.

Procedures of this nature were only just being established in many schools at the point when we ceased data collection so we would not wish to make any judgement on the basis of our existing evidence. However, one issue that emerged at this early stage concerned the wisdom of expecting teachers from schools other than the one applying for accreditation to sit on accreditation panels which were perceived to have an evaluative function. Whilst peer evaluation is a familiar concept in the literature of evaluation, our evidence suggest that schools which are unaccustomed to making self-evaluation public, have difficulty with this idea and feel that it might compromise future relationships between schools. Whilst we would not wish to deny the potential value of a collegial approach we feel that it assumes a level of openness for which some schools will need to be better prepared.

In this second main section of our across-site analysis of case studies we shift our focus of attention, from the substantive features of the projects we have been studying within schools, to a range of issues which broadly relate to the management of those initiatives.

Three qualifying points must be made about our remarks in this management section. First, it must be borne in mind that we are reporting on aspects of a **pilot** initiative. We fully acknowledge that schools have not considered their RoA projects to be provisional or experimental, but rather the beginning of what will become a permanent feature of their work. Nevertheless, the pilot status of their projects has imposed certain constraints, and permitted certain freedoms, which make it impossible to draw straightforward conclusions from their experience relating to a fully operating (national) RoA system. Only one of our case study schools, for example, had by the end of our fieldwork achieved what might be called a maintenance phase, and few had completed the transition from a project based on selected year groups and areas of the curriculum to a whole-school system.

Secondly, we have found it more appropriate to discuss the resource issues related to the management of RoA projects in the context of the various thematic sections, rather than in a separate section.

Thirdly, we have been keenly aware that the majority of management issues arising from the RoA projects are not unique to records of achievement but are instead general issues related to the management of change in schools. We have accordingly tried to confine our discussion here to those issues which we feel are specific to RoAs, and to acknowledge that our thinking in the area of RoA management has been tacitly informed by the wider body of research and scholarship in the area of innovation theory and the management of change. This body of knowledge is substantial, and its relevance and application to the management of RoA projects in educational schools deserves the most serious consideration.

2.1

Patterns of development and implementation

By way of introduction to our discussion of management patterns we would wish to emphasise that we are unable to delineate clear-cut models or types of management approach, especially ones based upon polarities such as 'top-down' and 'bottom-up', or 'school-based' and 'package'. In this section we shall therefore attempt merely to describe some of the more critical and RoA-specific **patterns** we have observed in schools' management of their projects.

2.1.1

Project starting point and dissemination

Case study schools have varied enormously according to the point or points in their complex structures at which they have decided to begin their projects. These project starting points have reflected a range of factors, including the character of the school and the sponsoring RoA scheme, as well as the personalities of staff who were felt to be most capable of establishing initial momentum.

(a) Target groups of pupils

For example, every one of the age groups in the secondary range is represented somewhere in our case study work as the age group which was initially targeted for RoA work. The principal consideration which appeared to have framed schools' decisions concerning which year group or year groups to target at the initial stage of development was either (a) whether to start with younger pupils, and then to phase the project in with new pupils over several years, or (b) to begin with older pupils, and then gradually to introduce the project to younger pupils, or (c) to attempt some combination of the two, including the introduction of the project to all year groups more or less simultaneously.

The main advantage of the first approach, i.e. starting with younger pupils, was that schools were able to capitalise on the generally less self-conscious and more enthusiastic attitudes, and greater impressionability of younger pupils when developing formative processes. Although it is probably too early to say whether this early start will make for a less nonchalant approach to the preparation of summative RoAs when these pupils are older, staff have reportedly enjoyed the opportunity to develop processes of recording without being constrained by the summative RoA. On the other hand, some schools which adopted this approach found that the lack of a clear conception of the summative RoA created confusion and a tendency to develop systems which were later found to be unnecessary or inappropriate. Whilst there was no wish to over-emphasise the importance of the summative RoA, it was somewhat difficult to pursue means without ends.

The second approach, i.e. starting with older pupils (and in particular fourth year pupils), had the clear advantage of enabling the production of summative RoAs within two years of beginning. This in many instances provided a great boost in that it demonstrated to staff and pupils what the end result comprised. In this way, perhaps ironically, the summative RoA may have been as much a contribution to positive attitudes as the processes of recording. As we indicated in 1.5.1, we have no evidence to suggest that pupils introduced to RoA-related work in their fourth year are necessarily better motivated towards that work than fourth year pupils who were introduced to recording processes as younger pupils.

It seems to us that the greatest problem with this approach is that schools were naturally tempted, having established a system of recording and reporting in the fourth and fifth year, to use a virtually identical system with first, second and third year pupils rather than to devise a system specifically aimed at the needs and abilities of younger pupils. However, this was not always the case, as the extract below illustrates:

Buttercup R.C. High School, Lancashire, had no choice in the target groups for the City and Guilds scheme. Although recognised as a useful starting point for development, aspects of the City and Guilds scheme such as the progress profile were regarded as inappropriate, and during the project the school expressed its desire to develop a scheme from year one that would be more suited to the school's needs, i.e. with the emphasis upon the formative process rather than the summative document. It was felt that the pupils at the school were at a greater advantage than most as far as finding employment was concerned, and so the emphasis on the summative document did not need to be as great. The school's own scheme was introduced into year one in 1987/88 with the intention that it would eventually displace the City and Guilds package at the top end of the school.

The third approach, which was to introduce the project to several, or indeed all year groups virtually simultaneously, seems to have been associated with an objective on the part of the school's management quickly to achieve an efficient, easy-to-administer whole-school system. This approach seems to have been less than successful, mainly due to the drain on resources of all kinds resulting from the attempt to develop processes and products at the same time.

It will be evident that we are unable to recommend any one of these three approaches without reservations, since we are conscious of their advantages as well as their drawbacks. Most schools' decisions, as to which of these approaches to adopt, will depend upon a range of factors unique to that school and not merely upon the strengths of one approach over another.

Similarly, different aspects of the curriculum have been targeted for initial work. Virtually all of our case study schools started their projects **either** through some kind of personal recording process from a tutorial vantage point, and principally involving form tutors, **or** through subject-based recording involving subject teachers. Given the likelihood that few schools will feel able to start on both fronts, however beneficial such an approach might be in an ideal world, we have no evidence to suggest that one approach is inherently superior to the other. In this sense we would advise schools to begin wherever they feel is most appropriate to their overall character, and wherever they feel that the prospects of meeting with staff enthusiasm are greatest.

(b) Curriculum starting points

It seems to us that there are similar dangers inherent in starting an RoA project on either side of the academic/pastoral divide. First, it can perpetuate that 'divide' at a time when many schools are examining ways of breaking it down, though this will depend very much on the philosophical underpinnings of the recording process as we indicated in the first part of our analysis.

Secondly, there is a danger that the initial curriculum starting point can, perhaps inadvertently, establish the character of the scheme in a way which may be difficult later to counteract however clearly the intention was to develop a comprehensive whole-school approach. This is probably as true of initially tutorial-based systems as they attempt to incorporate subject-specific elements as it is of subject-based systems which later attempt to incorporate a personal or tutorial element.

Thirdly, there is a danger that senior management teams may overrate the effectiveness of a dissemination strategy based on 'cross-fertilisation'. i.e. the assumption that staff making contact with RoA principles in one context will, with little effort or guidance, be able to adapt those principles to their work with pupils in a different context. This has perhaps been especially characteristic of schools which began their projects with form tutors and then encouraged those members of staff to explore ways of introducing RoA principles into their subject teaching.

> Throughout the pilot project at Columbine Middle School, OCEA, senior management placed considerable store by the theory that members of staff who made contact with RoA principles, in their capacity as form tutors operating the P Component, would gradually translate those principles into new academic curriculum practices in their capacity as subject teachers, thereby spreading the initiative through the school. However, members of staff admitted to finding this transfer difficult, not least because the practical steps for putting the principles underlying personal recording and reviewing into operation in the context of academic teaching have not been spelt out. It is also the case that the conventions associated with academic teaching can quickly override attempts at innovation which have little more than a philosophical basis.

Our evidence suggests that this kind of strategy is severely limited, as it depends very much on the initiative, and skills of curriculum analysis and development, of individual teachers working in isolation.

2.1.2
Management objectives

Variation in the general objectives of senior management in launching and sustaining a RoA project will be an important determinant of the overall development model. Our earlier remarks about the limitations imposed on our data by the pilot status of the projects we studied, are particularly relevant here. With a few notable exceptions, pilot schools voluntarily participated in their sponsoring RoA pilot schemes. One factor which must be accepted is that schools' participation in a RoA pilot scheme carried with it not only an obligation to carry out the work but also a certain amount of supplementary resources which were otherwise not available. Schools' reasons for opting–in will therefore only be relevant to schools in the post-national guidelines era in so far as they illuminate the kinds of objectives which pilot schools considered RoA-work capable of facilitating.

Among the many management objectives which lay behind case study schools' decision to participate in RoA pilot schemes were the following:

● The objective of fundamentally changing teachers' notions of 'achievement' as well as promoting new classroom practices which provide opportunities for pupils to develop, demonstrate and receive recognition for their achievement in this wider sense.

In a document prepared for the board of governors at Jasmine High School the then acting head teacher (who was later appointed head) stated that during the 1985/86 academic year the development of a policy of organised assessment of pupils would be a priority area. The Head's aim for the RoA project was to bring about a more positive approach to pupils' performance and to ensure that pupils received a meaningful document when they left school. He also hoped that the project would bring about changes in teaching style at the school, characterised by a less formal approach to teaching, individualised learning, and more problem-solving, experimentation and guided discovery.

Prior to the implementation of the pilot project at Sunflower High School the head teacher introduced a fifth year leavers' testimonial as a way of making teaching more efficient through the outlining of course objectives, bringing about a change in the traditional didactic teaching styles adopted by a large proportion of staff, and as a means of improving student motivation in the fourth and fifth year. Whilst the deputy head in charge of the curriculum did not regard student motivation as a problem lower down the school, the adoption of teaching styles inappropriate for the school population was a problem, and one of the reasons for seeking involvement in the pilot project was to bring about a change in this area as well as increasing the pace of curriculum development. The school had a high proportion of special needs pupils and only a small number of pupils who would meet success in GCSE, and the head felt that a strong emphasis on the academic side of the curriculum was inappropriate for such pupils, and that the school had to provide more opportunities for personal success with an emphasis on pupils' personal development. The RoA initiative was seen as a vehicle for bringing about this change.

● The objective of replacing one system for issuing reports to school leavers with another.

The head teacher at Begonia School commented that he had sought involvement in the RoA pilot project because he felt that the school should aim for a better system of reporting on pupils — one that gave a less subjective picture and with greater content than the traditional reporting system.

● The objective of increasing the motivation and self-esteem of 'disaffected' pupils.

At Geranium School the head teacher's primary aim for the RoA project was for it to act as a catalyst for curriculum change, bringing about increased pupil motivation through participation in the formative process and receiving a valued summative document in their fifth year. He also hoped that staff attention would become focused on learning objectives and outcomes and he emphasised his concern for the formative process rather than the summative document. His third aim for the project was that it would enhance the development of the partnership between staff, pupils, parents and employers in defining the school's aims and objectives.

The beginnings of interest in profiling at Campion Upper School, Suffolk, can be traced to the beginning of the decade when the school was involved with the Schools Council project. For some considerable time the school had been concerned about the low self-image of some of its pupils and senior staff had made serious attempts to introduce some form of personal profiling to give credit for non-academic achievement. Subsequently, the present Head became a member of the county working party and was involved in the early development of the Suffolk RPA. As had been the case at school level, and for the same reason, a deliberate decision was taken to make personal qualities a specific focus. Academic profiling was therefore never considered a priority as far as this particular development was concerned. When in 1985 Suffolk was included in the DES pilot schemes for records of achievement, and Campion Upper School was included in the Suffolk scheme, the initative was conceptualised as essentially contiguous with what had preceded it. Therefore the project at school level retained broadly the same objectives — to do with raising self-esteem through valuing personal qualities and achievements — as had been the case before.

● The objective of radically reorganising the school.

One of the most significant features of the project in Cornflower Special School, OCEA, was the Head's expectation that it would be the vehicle for reorientating the school in an educational direction and away from an excessive emphasis on medical welfare which, in his view, had led to unjustified acceptance of educational underachievement. He believed that involvement in OCEA would provide a rationale, strategy and all-important resources for introducing a subject-based curriculum and specialist teaching, as well as providing frameworks for assessment, recording and certification where none had previously existed. Thus the OCEA project and the reorganisation of the curriculum were inseparable and staff reaction to one was generalised to the other. Whilst some regarded these as exciting developments, others had deep reservations and regarded OCEA as a threat to those aspects of the school's approach which they had valued in the past. Whilst the morale of some staff rose, there were others whose morale plummeted. One or two staff also felt that a considerable number of less able pupils and their parents were deeply disturbed by the changes which had been brought about in the name of OCEA. The Head was inclined to interpret this as evidence that some parents, in particular, often wanted to take the line of least resistance and were unwilling to accept the realities of the outside world as long as their children were safely cocooned in the protective environment of a caring institution. Thus, despite opposition, he stuck to his guns and when opportunities arose he tried to impress on parents that the job of the school was to educate, and that OCEA was a way of helping more pupils to fulfil their educational potential.

The opportunity seized by the Head to acquire additional material resources should not, however, be underestimated as a management objective. In his view the previous administration had allowed the school to become seriously under-resourced and he explored all avenues, from jumble sales to ESGs, in his search for equipment, materials and staffing. OCEA was attractive in that it allowed schools to bid for what they identified as most needed in their context. Thus the Head's bid for equipment for a science laboratory, which had not previously existed, was accepted as a necessary prerequisite for science G teaching, assessment and recording.

● The objective of improving teacher-pupil relationships and the overall quality of tutorial/pastoral support in the school.

Whilst we would have to acknowledge that all of these objectives have been realised to a certain extent, their chief importance has perhaps been in the way that they have determined the overall character of the scheme and influenced the attitudes of staff. Expressing the purposes of the project in terms which incorporate the aspirations of staff may therefore serve not only to provide a selling point, but also to imbue the project with a wider significance which extends naturally from existing priorities.

2.1.3.
Strategies for involving staff

In view of the fact that most of our case study schools intended that their RoA systems should eventually be comprehensive and thereby involve all members of staff, it is not surprising that senior management teams have been preoccupied with gaining the support of possibly reluctant teachers. This task was attempted at a time when teachers' responsibilities were undergoing rapid and substantial change, and when teachers' union action in schools made it virtually impossible to rely upon traditional methods of recruiting staff to new projects. The concurrent introduction of GCSE must be singled out as perhaps the most notable force competing for teachers' commitment. Nevertheless, the **issues** arising from this are not particularly specific to records of achievement: they are issues which are regularly faced — and resolved — by senior management teams in schools when they have to mobilise staff support. The extensive literature on the management of change, has already delineated the kinds of strategies which can facilitate innovation in schools. The active and conspicuous support of the head teacher, the development of open, democratic styles of management and consultation have all been cited as essential. In this section therefore we describe some of the approaches which school managers have adopted in their efforts to gain the full and active support of staff specifically for the RoA project.

One strategy has been to devolve to heads of department, heads of year and to 'ordinary' classroom teachers, responsibility for the initial planning and conceptualisation of the project, and indeed for some day-to-day project decisions. This has in most cases been much more than a 'tactic' for involving staff: it has effectively distributed the work which would otherwise have to be shouldered by senior management. Moreover, many schools report that this kind of direct involvement in the **initial development** of their projects, on the part of classroom teachers, has been essential for the internalisation by staff of principles, commitment and effective practice.

At Columbine Middle School, OCEA, the pilot team of form tutors became not only a forum for the discussion of ideas arising from new forms of recording and reviewing, and a virtually autonomous decision-making body for the day-to-day management of the P Component pilot project in the school, but equally a curriculum development team, producing classroom materials and guidelines for teachers for a two-year tutorial course.

One way in which school managers have enabled this kind of initial participation by staff has been to convene working groups of various kinds. We have evidence that working groups of teachers have met for a variety of purposes: to consider matters relating to individual departments and year groups; to consider whole school matters; and to consider wider LEA and consortium matters. These working groups have been involved in planning and development, evaluation, INSET and staff development, and have met in school and outside, both during and after school hours, although teachers' union action placed serious constraints on the latter.

Schools have also varied in the kind and degree of consultation with staff which they have carried out in relation to their RoA projects. There has been a delicate balance to maintain here, since in some cases full consultation would almost certainly have brought about immediate rejection of the project by staff. In terms of schools' initial decision to participate in a RoA pilot scheme, some schools put the matter to a vote of the entire staff, while in others senior management took the unilateral decision to opt in.

Sunflower High School heard about the RoA pilot schemes via one of the regular county bulletins which invited schools to submit bids for inclusion in the project. Involvement was initially discussed at senior management level, and the head wrote to the LEA expressing interest in the project and asking for further information. After a visit by a county adviser, two staff meetings were held. The initial meeting was an information giving session at which the county RoA project co-ordinator addressed the staff. At the second meeting, a vote was taken and it was reported that the majority of staff were in favour of involvement.

At Buttercup R.C. High School only a few staff were involved in the decision to become involved in the pilot project. The City and Guilds scheme operated in years four and five and the initial group of five fourth year form tutors were asked if they would be involved in the scheme as reviewing tutors and if they would attend a county-based INSET course for reviewing tutors. Only two of the five form tutors agreed. Concern was expressed by uninvolved form tutors about the perceived lack of consultation on this matter, the feeling being that they were 'conscripted volunteers'.

Probably the aspect of consultation which was most crucial for the active understanding and acceptance of the project among staff was the quality of information which was made available, in terms of the nature of the scheme, the objectives on the part of senior management, the kind and degree of priority which senior management placed upon the project, the role which staff were expected to play and the ways in which the project related to their existing practice and principles. This quality of information often determined whether staff perceived the project as essentially democratic or autocratic, and whether consultation was genuine or cosmetic.

At Hydrangea School consultation during development was facilitated by regular 'Assessment and Profiling' bulletins, in which details of the project could be described. Although most articles in the bulletins were written by the school RoA co-ordinator or by the ILEA Development Officer assigned to the school, the bulletins at times served as a forum within which the viewpoints of individuals and small groups could be made available to the whole staff.

2.1.4
Uniformity of approach across the school

Development patterns also differed according to the degree of uniformity across the various elements of the project. This kind of diversity has characterised, for example, recording documents prepared by different departments, approaches to the recording process by different members of staff, and development strategies adopted by different departmental or tutorial teams.

Whilst it is true that individual departments at Bluebell School carried out their profile development work within shared principles and objectives, it is also true that the RoA project at Bluebell is perhaps best understood as a network of individual departmental projects, each with its own objectives, terms of reference, pace of development and working style. The continual change and adaptation to which work across departments was subjected, and the variability of work across departments, was for the most part a planned and valued feature of the project, which was never envisaged as an entity which would at some point eventually assume some kind of static unity. The upshot of this autonomous model of development was that a very wide range of approaches co-existed within the same school. Suggestions that the format and content of subject profiles should have been standardised, which were sometimes made by the Head, and occasionally by year heads who found the lack of uniform timetable for issuing subject profiles an irritant, were resisted by staff. The very wide range of profile documents at Bluebell signified not merely a surface diversity in format and presentation, but also a diversity in underlying assessment principles.

The RoA system at Daffodil Comprehensive School, Wales, was by design exceedingly uniform, both within the school and across the eight Gwent pilot schools. At the consortium level, the eight schools were considered to be employing sufficiently uniform materials and methods to be treated as a single scheme in the WJEC monitoring exercise. At the school level, all teachers administering the system used identical materials. There was some variation in the way that the subject panels defined their task in the development of comment banks, but these variations were mostly reconciled by the common comment bank format and by the commonality of approach within each department. In terms of the administration by form tutors of the Record of Personal and Social Development, some variations were observed in the way that tutors organised their interviews and in the way that they mediated the meaning of statements and the nature of the exercise to pupils.

Although departments at Geranium School followed a common timetable for development of RoA schemes, they were left to decide upon the details in terms of: the identification of assessment criteria for first year pupils, the development of a programme of appropriate assessment techniques, the organization and frequency of assessment, and proposals for recording and reporting formats. This led to a diversity of approach across the school.

Whilst a considerable degree of diversity may be essential in order to accommodate wide differences of philosophy and practice within the school, and to reinforce the 'bottom-up' flavour of some schemes, some schools may feel the need to ensure a degree of uniformity to facilitate acceptability and clarity. Moreover, in a dissemination phase, as opposed to a pilot phase, this need may become more prominent if ownership of the project comes to reside less in the school and more in the scheme. Accreditation **principles** which will form part of national guidelines may be sufficiently general to accommodate considerable diversity of approach within a school, but accreditation **procedures** will almost certainly treat schools as monolithic and expect them to have internal mechanisms to ensure a substantial degree of within-school uniformity. The issue for the future is therefore not resolved by pilot experience.

2.2 Special roles and responsibilities

In this section we examine issues surrounding the contribution made by individuals and groups who had special roles in relation to the development and implementation of RoA systems in case study schools. First we look at the role of school co-ordinators: secondly we discuss the role of ancillaries; and finally we describe the contribution of parent, employer or community groups. We do not examine here the contribution of individuals and groups with designated roles in the conventional organisation of schools, such as headteachers, deputy headteachers, heads of department and classroom teachers, because we have discussed their input at various points throughout this report. Here we are more concerned with roles which have been specially created or extended in response to the perceived needs of the RoA initiative.

School co-ordinators played an important part in RoA projects although, not surprisingly, they had a less prominent role in schools where an RoA system was up-and-running early in the pilot phase, where development involved substantial delegation, or where development officers or advisory teachers who were appointed at scheme level were closely involved in development processes within schools. In general, however, projects have depended crucially upon the effectiveness of school co-ordinators in mobilising members of staff, in liaising between a wide range of staff within and outside the school, and in fostering enthusiasm, new attitudes and new practices.

In view of this substantial responsibility, it is not surprising that, in the great majority of our case studies, school RoA co-ordinators were members of the senior management team or were appointed to the senior management team in their capacity as RoA co-ordinator.

(a) Position of the school co-ordinator and implications for organisation

The school profiling co-ordinator at Bluebell School was, before her appointment, head of the English department and already considerably knowledgeable in profiling matters in the context of her own department. She was promoted to the post of Senior Teacher as part of her appointment as school RoA co-ordinator. Her appointment reflected the view on the part of senior management that she was widely liked and respected by staff, but not directly associated with senior management, thus reinforcing the teacher-led model of development which the senior management team wished to adopt. It was hoped that the co-ordinator would be able to mobilise and sustain enthusiasm, commitment and industry, in a way which senior managers may not have been able to do, especially during the dispute. This strategy was generally felt to have been successful, although there was a certain amount of ambiguity in the roles of the co-ordinator and the deputy head with responsibility for pastoral matters, in relation to the tutorial side of the project.

Where co-ordinators were not senior managers, problems sometimes arose at the point when decisions were required regarding institutional policy. During initial development work co-ordinators who were more junior were often able to generate commitment and enthusiasm from their peers, but there inevitably came a time when development work had to be translated into policy. Co-ordinators who lacked a role in the central decision-making processes of the institution discovered that much of their time and energy was spent lobbying those who could exert more leverage on the system.

Nasturtium Tertiary College chose to appoint two co-ordinators who worked together as a team. One was appointed from outside the college to a Senior Lecturer post funded for three years. The other was an L2 maths lecturer who was internally seconded to the project and given an SL for the duration. Within the management structures of the college these were relatively junior posts, and co-ordinators experienced some difficulty in getting the policy decisions at whole college level that they regarded as necessary. They had backing from the Principal and a Vice Principal but they had no real power to formulate policy themselves, thus they spent much time lobbying senior and middle management who often had only a tenuous grasp of the issues and did not appreciate the urgency if a system was to be in place before co-ordinators' contracts ran out.

Until Easter 1987, the school co-ordinator at Geranium School held the position of head of year. After this time he became a deputy head with responsibility for first and second year, records of achievement and some administration. As a head of year he had felt that it was important for him to be given extra 'clout' in the form of the deputy head's signature on documents sent round to staff. He also found that he had to negotiate with the headteacher and the deputy in charge of curriculum in order to get decisions made at senior management level for the changes he wanted to implement. There were no problems with this; he felt he got through the changes he wanted.

On balance our judgement is that RoAs should be co-ordinated by a senior manager with whole-school responsibilities, both during development and possibly in the maintenance phase when co-ordination might amount to little more than monitoring and oversight of an established system. However, we recognise that this suggestion has implications for the way in which senior managers may be expected to distribute their time.

During the pilot phase, some special timetable provision was made for the work of school co-ordinators in some case study schools. However there was little consistency across the case study schools in the amount of non-contact time which school co-ordinators were allocated specifically for RoA work. Indeed some co-ordinators had no extra non-contact time, others had a notional half a day or a whole day per week, and others were allocated time equivalent to 1.3 extra staff on the basic pay scale. This last arrangement effectively freed co-ordinators for RoA work for 50–75 per cent of their time and enabled them to have some additional co-ordination support. In general the amount of non-contact time made available was related to the size of the institution but more particularly to the amount of development work which was expected to be carried out at school-level. Timetable provision was noticeably less in schools within schemes in which much of the initial development work was already accomplished at scheme-level.

Because of the severely limited level of funding available to each individual pilot school in Gwent, the Management Panel of pilot school heads took the decision to channel what staffing resources were available to cover staff taking part in e.g. consortium panel meetings rather than to free school co-ordinators from teaching commitments. As a result, none of the Gwent school co-ordinators had any additional non-contact time allocated to them for RoA-related work. The school co-ordinator at Daffodil Comprehensive School, Wales, was therefore required to carry out RoA work either in his own time or in his existing allocation of time off timetable. He was, however, able to use resources for staff cover to enable him to attend meetings outside school.

During 1985/86, the school co-ordinator at Buttercup R.C. High School was allocated half a day per week for RoA-related activities, but as this was not timetabled she did not take it. In April 1987 she was teaching for 28 periods a week and was given approximately two hours per week for the management of the RoA project. The management of the scheme included, amongst other things: organising supply cover for reviewing tutors; organising the pupils and tutors for reviews; providing reports for governors, academic/pastoral board meetings; talking to the local Rotary Club, press, teachers and other schools; supervision of a student carrying out an investigation of RoAs from a local university; evaluation of the project as part of the county project local evaluation; supervising the computing and running word-processor training courses for tutors; and carrying out INSET in other LEA schools. During 1987/88 she and indeed all the Lancashire school co-ordinators were involved in the provision of INSET for other county schools on a more regular basis. This commitment took her out of Buttercup for an average of 2–3 days per week. This made the management of the RoA initiative within the school very difficult.

The school co-ordinator at Poppy County High School was allocated 25 non-contact periods (out of a total of 40) per week specifically for RoA work. This was also supplemented by some time for assistance from a more junior member of staff. Altogether the ESG grant provided for the equivalent of 1.3 extra members of teaching staff on the basic scale. In relative terms this was generous support although the first co-ordinator felt that the project was in fact under-resourced in terms of the whole school system which she established. One reason for this probably had much to do with the fact that the system in the early days was heavily materials dependent and centrally controlled. The sheer administration of the project in terms of dealing with the bits of paper therefore consumed a great deal of co-ordination time. This noticeably diminished as a devolved system took its place and the strict adherence to comment banks was abandoned.

Other commitments did not encroach on the time of the first co-ordinator as much as might have been the case had she been a deputy head. However, she retained some responsibility for departmental matters for at least the first year and these eroded some of the time she had for the RoA project.

As some of these examples illustrate, school co-ordinators who were also senior managers experienced difficulty in taking their time allocation where it was available, or fitting in RoA work where it was not. For the most part they were not relieved of their existing responsibilities in order to devote time to the RoA initiative, therefore conflicting demands on their time created serious tensions. Although the time conflicts which characterised the start-up period may become a little less problematic when the project enters a maintenance phase, the oversight or monitoring role is still likely to be substantial especially when the whole school is involved. Senior staff will clearly not be able simply to add RoA co-ordination to their already substantial job descriptions and time commitments and other strategies will need to be sought to resolve this difficulty. In one of the local authorities in which some of our case study schools are located a decision has been taken to increase the number of deputy heads in each school from two to three in order to accommodate all the extra administrative work occasioned by a number of new initiatives.

There was considerable variation in the amount of previous experience which school co-ordinators in our case study schools had gained of records of achievement and related areas, before formally assuming their duties in the context of the pilot scheme.

(b) Background and experience

Despite the fact that the school co-ordinator at Daisy High School was an established member of senior management, and that she had been involved in the production of the school's previous 16+ leavers' profile, her main feeling when she took on the task of RoA co-ordinator was that she had no prior experience in or knowledge of the area of RoAs upon which to draw, and she reported her need to 'learn quickly'.

Previously the deputy in charge of the Lower School, the school co-ordinator at Rose School had put much effort into developing strong tutorial teams and pastoral programmes based on Active Tutorial Work. By 1985, pastoral work of this kind had been used for three years with years 1, 2 and 3, and years 4 and 5 and the sixth form were engaged in something similar. She was however very concerned with the need to break down pastoral/academic barriers and perceived that records of achievement might be a vehicle.

Our evidence suggests that school co-ordinators' lack of specific previous experience in profiling or records of achievement was not in itself a disadvantage. Certainly they were burdened initially with the task of quickly becoming experts, but most were accustomed to this in the normal course of their work. Effective leadership styles and the ability to absorb new knowledge and principles quickly would therefore appear to count for more than previous RoA-related experience in itself.

However, it would appear from our evidence that other forms of previous experience such as curriculum development and/or staff development were a distinct advantage.

The backgrounds of both co-ordinators at Nasturtium Tertiary College had a bearing on their appointments to this work. One had wide experience in staff development in both education and the sevice industries. Senior Management clearly saw staff development as the RoA co-ordinators' most crucial role although they also seized the opportunity of the appointment to strengthen this aspect of the college in other ways. The other co-ordinator had previously been head of a sixth form and was able to develop and build on his experience of working with staff from across the curriculum.

The OCEA school co-ordinator at Wisteria School had for some time previously been promoting and supporting similar kinds of changes within the school, through her school-wide curriculum development responsibilities and in particular through her involvement with such projects as the DES Special Project (LAPP), the development of a tutorial programme throughout the school, and so on.

Also with regard to background and experience, it was of some interest that in those schools where the RoA project was co-ordinated by a deputy head, the specialist responsibility of that person tended to reflect the dominant flavour of the project in its intial stages — though which was cause and which was effect would be difficult to assess. However, when attempts were made to broaden the scope of the initiative in terms of curricular coverage, co-ordinators responded by consciously extending their definitions of their area of responsibility, by delegating some of the work to others with more specialist experience, or by handing over co-ordination to someone whose job description better fitted the direction of the project.

(c) Mode of working

Our evidence suggests that school co-ordinators' main mode of working, certainly in the early stages of the pilot scheme, was through personal approaches to individual members of staff, usually heads of department and heads of year. This was, of course, especially necessary during the prolonged period of teachers' union action — which in some instances lasted until December 1986 — when it was not possible to hold meetings out of school hours. When union action was at an end, approaches to individuals remained important although a greater diversity in modes of working became more evident. In various proportions co-ordinators tended to combine the use of memoranda and discussion papers, informal personal contracts, small group and whole staff meetings and delegation. The relative prominence given to any one of these elements depended upon structural constraints (e.g. the size of the school), established habits (perhaps stimulated by union action), institutional norms and, most crucially, personal style.

Given the kinds of factors which make one mode of working more feasible in one setting than another we would not wish to make a judgement about which approach is to be preferred. However, we would wish to emphasise that whichever mode is deemed appropriate, it carries with it implications both for the appointment of suitable staff, in terms of expertise and aptitude, and for resource support, especially in terms of time allocation.

We have evidence that where schools took on ancillary staff to assist with the RoA pilot project, their support was felt to be important at the time.

During the pilot phase at Poppy County High School the school was able to employ a part-time clerical assistant to support the work of the project for twelve and a half hours per week. During the first phase her services were indispensible because the comment bank system demanded a vast amount of typing, reprographing, collation and distribution of formative documents and electronic storage and retrieval of data for summative documents. In order to produce 250 summative documents of approximately 35 statements each, the assistant had to work full-time for six whole weeks. This was recognised as another reason why the system was not really viable. Under the new arrangements introduced in autumn 1987, it was decided that summative statements should be typed by the clerical assistant whilst ESG support was still forthcoming. It was not clear what the situation would be in the future. Three possibilities suggested themselves: resources for continued secretarial support might be found; statements might be prepared on computer by pupils and/or teaching staff and printed out on single sheet format; or statements could be hand-written. Style of presentation was however felt to be important especially since the school had to reverse any poor impression that the first issue of statements had created.

At Cornflower Special School, the OCEA assistant's role was central. Although technically appointed as a clerical assistant, this person was the focus of all P Component recording from 1985-87. It was she who encouraged children to make their own records. She spent time in lessons recording her observations of children's achievement and then conducting one-to-one review sessions and producing interim summary statements with the pupils two or three times a year. She also produced summative records for this personal recording element, and co-ordinated production of curricular statements. She therefore did much of the day-to-day co-ordination of RoA work in the school, although issues of policy remained the province of the head and the school co-ordinator. It was significant that after her contract came to an end the amount and quality of personal recording declined. This served to reinforce the Head's perception that additional assistance of this kind was vital to the success of the initiative. The degree of responsibility carried by the ancillary in this situation was not entirely without precedent. It was related to the fact that the conventions operating in special schools differ from those in ordinary schools, and ancillary staff, such as nurses, work closely with teachers as part of a unified team. However, for a clerical assistant to take on such a role must be regarded as unusual and it appeared to exacerbate the difficulty of finding a teacher to take on the role after her departure.

In other schools there did not appear to be the same need for additional ancillary help. Administrative tasks were accomplished by giving an extra time-allocation to existing teaching and secretarial staff, or by enlisting the help of central clerical staff attached to the project team at county or scheme level.

In the long-term the argument for ancillary support revolves mainly around the means of production of final summative statements. However, as we have stressed throughout this discussion of management issues, long-term production needs are difficult to calculate from our evidence because only one of our case study schools had reached steady state by the close of data collection. When summative documents are being produced for all fifth year pupils, and all other pupils are fully involved in formative recording across the whole educational programme, the extra burden on existing staff may well require the employment of additional staff. This is likely to be particularly true if a decision is taken to present documents in typed script. Handwritten documents are unlikely to make the same demands.

Interestingly we encountered different opinions of the likely need for ancillary support where computer-based systems were developed. Where computers were extensively used, or were planned to be used, they were expected to save time eventually. But the time saved was conceived mainly as teacher-time and it was anticipated that there would be some continuing need for specialist support for the storage and retrieval of data and the maintenance of computer systems.

2.2.3
Support role of parents, governors and the local community

Within some of our case study schools representatives of parent, governor and community groups, including employers, were involved in consultations during the pilot phase.

> At Jasmine High School several parents meetings were held to discuss the RoA project, and a number of parents volunteered to take part in three parent/staff sub-groups established to examine pertinent issues. Over forty parents volunteered for this aspect of the project, and there were representatives for each year group as well as parents of pupils at Jasmine's feeder primary schools. It was felt that a significant contribution to the meetings had been made by those parents who were employed in industry. The governing body at Jasmine was also very supportive of the RoA initiative. All were invited to weekend conferences where RoAs and other curriculum initiatives were discussed. Representatives were also invited to a whole-school INSET session held in February 1987.

In this example, as in similar cases, the boundary was blurred between, on the one hand, direct and active involvement in development and on the other hand, more conventional and general exchange of ideas with no obligation on either side to act. For the most part the involvement of outside groups was confined to the latter category although the opportunity for schools to explain their project to potential users, and for users to express their views at an early stage was undoubtedly valuable. Evidence for the alternative was scarce but where individual representatives of groups had been drawn more intimately into the development process, a certain feeling of tokenism was almost inevitable. On both counts a particular difficulty was, as always, the temptation to involve articulate, middle-class parents and the big local employers who might in the end be seriously unrepresentative of user groups.

2.3
Relationship with LEA and consortium schemes

The nature of the relationship between the RoA project in our case study schools and the local consortium project team varied considerably according to the patterns of development promoted by schemes (see 2.1). For example, where substantial elements of the project were developed at scheme level, external input was mostly in the nature of support for implementation, especially through in-service training, and a degree of quality control.

> Individual members of the Suffolk LEA team tended to make one-off appearances at Hawthorn R.C. Upper School when responding to requests from the school for in-service sessions. Thus a day-long in-service session was conducted with the fourth year tutor team in 1985/86 and another with the same group as fifth year tutors in 1986/87. Although centrally-convened, school co-ordinators' meetings were organised, it was the general policy of the county team to respond to requests for support. This policy was itself a practical response to the fact that the sheer number of schools involved in the pilot scheme (39), and the consequent limitations on the resources available to each school, meant that time and effort had to be spread thinly. Another means of county support, also a consequence of the difficulty of working closely with so many pilot schools, was through publication of materials. In the course of the pilot, groups and individuals connected with the county team produced a handbook for school co-ordinators, a substantial report on reporting systems, and a report on profiling cross-curricular skills. These were regarded as having

considerable value and the cross-curricular sub-group in Hawthorn School made particular use of the cross-curricular skills report in its discussions. However, publication as strategy for development assumes a long time-scale so little impact in terms of implementation was discernible by the end of the pilot phase. Indeed the cross-curricular skills report only became available in February 1988.

It is a striking characteristic of the OCEA scheme that LEA consortium support was much in evidence at Periwinkle School. This is essentially the other side of the coin from a good deal of external imposition in the nature of the scheme. From the earliest stages of the pilot study, teachers were seconded in each of the G Components and in P to work for one day a week at Westminster College, Oxford alongside their equivalents from other pilot schools. Activities included: general INSET lectures, working on the development of criteria, as well as individual course work at the Oxford University Department of Educational Studies. An OCEA resource centre was originally set up at Westminster College, this has now moved to a central county centre in which resources for all the new county initiatives are gathered together under the control of a senior adviser. During 1986/7 a Directory of Practice in Oxfordshire Schools was developed in which individual schools contributed information about particular aspects of development work. These were then filed in the Directory of Practice for the benefit of other schools. Provision is made for new co-ordinators coming into the scheme to meet existing co-ordinators to learn from their experience and from head teachers. There is also an Oxfordshire OCEA Newsletter entitled *Reynold's News* which keeps various members of the pilot scheme in touch with one another and with ongoing developments. There are also general OCEA newsletters and guides have recently been published for the P Component and for each G Component. There has clearly been active personal support from the LEA in terms of both resources and personnel. By contrast the support received directly from the Oxford Delegacy has been minimal. In some cases where the school has asked the Delegacy what it should be doing the response has been, "You tell us." At Periwinkle the co-ordinator felt that in OCEA a partnership between LEAs and the Exam Board has been pushed to the limit. As the county moves into a new development mode in which the range of curriculum and assessment initiatives in the county are brought together under the GONOT umbrella, so this support is likely to be more LEA focused and cross-initiative. In future provision is to be made in school for a co-ordinator on a half-time secondment to oversee and integrate the development of new initiatives within the school. OCEA is becoming increasingly a school-based system. LEAs are increasingly divergent in their practice. For the school co-ordinator the consortium level network has little reality; the co-ordinators of schools in the four counties having only met twice in the three years of the pilot scheme.

On the other hand, where there was a greater emphasis on school-focused development of RoA systems, perhaps based on some statement of general scheme principles or criteria, the contribution of the county or consortium team was diverse. In some cases local co-ordinators, or development officers or advisory teachers attached to specific schools, provided regular and substantial on-site support for development.

Hydrangea School, had the benefit of a Development Officer from the ILEA team on site for one full day each week. This development officer worked closely with the school RoA co-ordinator, heads of department and faculty, heads of year, and individual classroom teachers, feeding information and suggestions, preparing materials, and providing the school with regular oral and written accounts of the project's progress. Senior and middle managers in the school emphasised the difficulty they would have faced in moving quickly through the start–up phase without this regular support and involvement.

In other instances, school-level development was conducted with comparatively little direct assistance from the LEA or consortium.

The RoA project at Daisy High School, Wales, was based on the in-house production of subject and personal comment banks. The project was conducted with comparatively little direct assistance either from the consortium, due to the very large number of schools involved in the Welsh scheme and to its comparatively low level of funding, or from the LEA. The principal outside reference point was the general criteria issued by the WJEC, and it was left largely to the school to interpret those criteria and to devise ways of satisfying them.

The perceived lack of support from the LEA was an important issue in the early months of the City and Guilds project in both case study schools in Lancashire. The project co-ordinator and the four school co-ordinators felt very isolated within the authority, receiving no support from the advisory service, although they were present at LEA-based INSET courses. Later on in the project the situation appeared to improve, with adviser support for the initiative being more obvious. For example, they made visits to the Lancashire pilot schools initially to gather information it seemed, but later to offer assistance. One member of the advisory service formed part of a panel of speakers at a parents' evening held at Buttercup R.C. High School for the parents of the initial cohort of fourth year pupils. However, at Sunflower High School the advisers provided encouragement rather than expertise. The school co-ordinator felt that the provision of expertise had been left to the four Lancashire school co-ordinators and the Lancashire project co-ordinator. A county working group was established to examine the matter of a Lancashire RoA policy.

It needs to be said however that, irrespective of whether the initial development was at scheme or school level, the nature and degree of contact between school and scheme project teams was also determined to some extent by the level of funding in relation to the number of institutions involved in the pilot phase. For example, there were a large number of schools in the pilot in both Suffolk and Wales which meant that resources (including human resources) available to each school, were very limited.

In many cases the relationship between the school and the scheme was harmonious. There were some cases, however, in which both sides admitted considerable tension or even friction. Where the relationship was harmonious the school either appeared to share the same commitment to the same principles and processes as did the LEA or consortium team, or the scheme allowed the school substantial freedom to develop and implement the RoA system as it saw fit as long as it fulfilled certain base-line criteria. Where the relationship was more difficult, tensions seemed most often to arise over different interpretations of what was implied in the notion of piloting the initiative, and in particular how pilot work was expected to contribute to school and scheme RoA policy.

As we pointed out in 2.1, it would be too simplistic to characterise development as conforming either to a 'top-down' or a 'bottom-up' model. No scheme developed a package that was expected to be adopted by schools without further development or modification; neither did any scheme allow schools total freedom to develop school-based approaches without reference to scheme principles. All development involved some input from both school and scheme and in many cases the relationship was expected to involve interaction and dialogue. Indeed some schemes had an expectation that initial ideas would be developed and refined by experimental work in schools but that the ultimate objective was to establish policy at scheme level. In terms of models of innovation and change such an approach could be described as centre-periphery-centre. Whereas this approach has much to recommend it in theory — it may even appear ideal — our evidence indicates that some schools experienced difficulty in holding on to the idea that they were participating in scheme-wide piloting work which had an experimental aspect. It is part of the conventional ethic of schools that pupils should not, as far as possible, be used as guinea-pigs and this meant that some teachers and institutions were unsympathetic to the prospect of a process of trial and revision which did not move fairly rapidly towards steady state in the form of policy.

Two reactions to the inherent ambiguity of this kind of situation were evident from our case study work. On the one hand school staff sometimes became committed to ideas which they were not altogether happy to abandon in order to fall in line with later thinking at scheme level.

In addition to providing substantial financial resources, the county project team made regular contact with Poppy County High School, Essex: one member was attached to the school and expected to visit once or twice a week to assist with development work. County inspectors also made frequent visits and staff were invited to participate in the six county working parties which were set up to explore particular themes (although only one was attended by one member of the school's staff). On the whole the relationship between the school and county project teams was not harmonious. The county team wanted to encourage school-focused development based on a set of agreed principles before drawing on this experience to formulate county policy. However, Poppy County High was initially highly committed to the comment bank system which it had decided upon prior to the commencement of the pilot phase and it was reluctant to consider radical alternatives. Whilst the county welcomed the whole-school and whole-curriculum approach and the central role of one-to-one teacher-pupil discussion, it was unhappy about the lack of personal recording and the dependence on tick-box type comment banks. It was also unhappy with the school's apparent preference for ploughing its own furrow rather than contributing fully to the development of a county scheme. The county team itself had a fairly strong leadership style but it made relatively little impression on developments in the school at this time. Major changes in the RoA system had to await other changes within the school.

These came after the departure from the school of those chiefly involved in the setting up of the existing RoA system, and following the largely negative response to the first summative documents. At this point, the school was able to reappraise the whole project. The changes that followed were dramatic and much welcomed by the county team because they were more in line with developments elsewhere. A member of the county team began to work in the school once more and the new school co-ordinator was drawn into the county's plans for dissemination by being made responsible for county in-service in relation to recording in a particular curriculum area.

On the other hand, some case study school staff expressed irritation that they had too little guidance from county and consortium teams about the kind of end-point in terms of product or policy they were expected to be working towards.

External support at Magnolia Girls' Grammar School, EMRAP, was provided by the LEA co-ordinator who visited the school several times, who was in telephone contact and who arranged regular meetings for all school co-ordinators in the town. In addition the regional i.e. consortium co-ordinator visited the school on two occasions and had contact with school personnel at annual regional conferences. However, all parties admitted that the relationship had not been easy and this was reflected in the different perceptions that the school and county teams had of events. For example, the local co-ordinator said that the school had received concrete guidance and support, including sample materials, from the local and consortium project teams from the very beginning. However, in his view, the school showed an unwillingness to be fully involved at local and consortium level and some representatives had walked out of early consortium events. He claimed that even at the end of the project the school had not requested INSET support and, unlike other schools, had had no in-service sessions with county inspectors, advisors or the co-ordinator. Although the school co-ordinator had two lengthy meetings with the county inspector, and the decision to ask parents to make a contribution as part of the formative process came from one of these meetings, the school team felt that the general guidance received from the county and consortium teams was nebulous and examples of materials from other projects were of little use without some clear conception of the desired end-product. Staff felt that advice concentrated on processes without giving them an end in view. They argued that they could not consider

appropriate means until they had an idea of where they were going and they were not convinced that processes should be ends in themselves. Staff also reacted negatively to what they perceived as evangelism and implicit critism of their current practice. Towards the end of the pilot phase relationships appeared less strained although there were further misunderstandings along the way. For instance, it was the school co-ordinator's understanding that experimental school-based development was to be a preliminary to agreement on a common LEA system. It was only later that she realised that schools were expected to develop their own systems provided they satisfied the principles on which accreditation was based and provided the format of summative documents was roughly similar. On the other hand the LEA co-ordinator claimed that this had been made clear to schools in the pilot as early as September 1985. It seemed that most of these differences in perspective arose out of fundamental differences in commitment and motivation attributable to the factors surrounding the way in which the school came to be involved in the project in the first place.

In essence these two kinds of reactions were opposite sides of the same coin and attest to the delicate balance between freedom and control that has to be achieved in approaches to piloting which aspire to be interactive, and essentially democratic. Whether these observations have relevance for development in the post-pilot phase will, of course, depend on the nature of national guidelines and decisions about the management of local schemes. If, as seems likely, it appears necessary for schools which are as yet uninvolved to go through some of the same development processes, then these are issues which will need to be addressed in the planning of appropriate strategies.

2.4

Changes in curriculum, teaching and organization

In this penultimate section of our across-site report, and in the one which follows, we shift our focus once again; instead of simply reporting on our observations during the pilot phase (the future relevance of which may be somewhat limited), we now turn our attention to what we perceive to be emergent trends of considerable future significance. Such trends may not have been fully realised in practice in our pilot schools, but our comments nevertheless relate to aspects of development which have at least been partially addressed. In other words our analysis is grounded in the data but sometimes goes beyond the information given. Our aim is principally to draw together a number of strands of development.

The third of the four purposes of records of achievement in the DES statement of policy relates to the need for schools to assess the extent to which their curriculum and organization actually provide opportunities for pupils to develop and demonstrate the skills and abilities which the new processes of recording are to document. Schools setting out to produce an all-round picture of a pupil should 'consider how well their curriculum, teaching and organization enable pupils to develop the general, practical and social skills which are to be recorded' (paragraph 11). Many schools probably feel that they already aim to help pupils to develop their all-round potential. If this aim were already genuinely being achieved, the development of records of achievement would involve no more than identifying what the school is already fostering, and enshrining that in a recording system. In reality the picture is not so straightforward. First, records of achievement are widely seen as a catalyst for re-appraising the school's aims as a starting point for structural and curricular change. And secondly there is no clear distinction between assessment-led changes to curriculum and curriculum-led changes to assessment, since any examination of aims, curriculum, pedagogy and recording is interactive.

We have at several points in this report discussed the substantive issues relating to this problematic relationship between recording objectives and learning opportunities. In this section we present evidence relating to more structural changes to curriculum, teaching and organization, and draw out some of the management implications of those changes.

2.4.1
Curricular provision

A recurrent theme in our evidence is the fundamental nature of the relationship between the recording of achievement and the provision of curricular opportunities. At several points we have referred to the fact that an RoA system can be considered invalid if it does not reflect the educational experience of pupils. More specifically, opportunities need to be provided for pupils to demonstrate various aspects of achievement if it is expected that such achievements will be assessed and recorded. Indeed it contravenes natural justice if pupils are assessed in areas where they have no equal access to opportunities for achievement. And this applies as much to social and personal skills and qualities as it does to achievement in academic subjects.

During our investigations out attention was drawn to ways in which the development of records of achievement has gone hand in hand with curriculum review and development, although it has been difficult to disentangle cause and effect and the influence of other concurrent initiatives. Nevertheless, our evidence suggests that many case study schools have begun to reconsider their extra-curricular, pastoral and academic programmes in this light, and the relationships among these elements of their total curricular provision.

For example, all of the schools we studied developed a personal element as part of their new recording procedures, whether or not they had a pastoral curriculum. It was also the case that most of our schools saw personal recording of extra-curricular experiences and achievements as belonging to the pastoral side of their organization. Beyond this commonality, however, lay considerable diversity in the way schools defined the relationship between recording of personal achievement and their pastoral curriculum. At the most general level, we were able to distinguish between those schools that did, and did not, have explicit pastoral programmes before the introduction of a personal element in the record of achievement. If we include such courses as 'Life Skills', 'Counselling', 'Active Tutorial Work', 'Guidance', 'Personal Development', 'Personal and Social Education' and 'Careers' under the pastoral rubric, very few schools fell within the latter category. Where a pastoral programme of some description existed previously, the personal element of recording was usually assigned to some part of it. In some of these cases the personal element of recording was simply added to the programme, with no real consideration given to whether the content or methods used provided opportunities for pupils to develop and demonstrate the skills and qualities being recorded. In other cases, however, pastoral programmes were scutinised in order to root the new personal recording element in school or classroom activity.

Records of Achievement

Where no pastoral programme existed previously, the personal element of recording tended either to be assigned to form teachers in registration periods, where it had little impact on the curriculum, or it was the catalyst for substantial development of new pastoral structures and programmes.

Two stated aims of the RoA project at Margiold High School, were (a) to foster a positive tutorial function which was seen as weak, thereby changing the role of form tutors from that of markers of registers to that of monitors of achievement, and (b) to highlight the pastoral programme that was needed by the school. There had been problems in the past with form tutors being wasted and pastoral schemes which had developed gathering dust. Indeed the head of first year was concerned that this might happen to RoAs.

Continuous personal recording was assigned to registration time which, for administrative reasons, was increased from one period per week to four periods per week for pupils in **all** year groups and not only those involved in RoA activities. Tutorial programmes were developed by all year teams during 1986/87 and were seen as important in providing pupils with the opportunity to develop and achieve in the pastoral arena. In years two and five the pastoral RoA was developed alongside the tutorial programmes to ensure a close fit between the two. However, there seem to have been problems of mismatch in year one, where the RoA was devised by the RoA steering committee and the tutorial programme by the first year pastoral team. It was reported in 1987 that the first year pastoral RoA was still in the hands of the steering committee leading to a division of responsibility not apparent with other years. It was felt that there was only a very tenuous link between personal recording by pupils in year one and the tutorial programme, in that pupils kept worksheets completed as part of the latter in their personal interests and achievement folders.

This evidence indicates that consideration must be given to the structural and substantive relationship between personal elements of recording and pastoral programmes within the curriculum.

First, there is the question of whether the pastoral side is the most appropriate place to assign the personal element. There may be a risk that this will serve to reinforce the divide between academic and pastoral curricular coverage, and prevent discussion of ways in which the two might be more closely integrated. There is therefore good reason to explore ways in which the personal development aspects of teaching and learning within the academic curriculum can be made more explicit (see also 1.1).

Secondly, there is the allied question of how dependent personal recording elements should be upon special personally-orientated classroom work. On the one hand, form tutors interviewed felt that a personal element considered separately from other forms of curricular recording was difficult to operate validly. On the other hand, the greater the dependence upon new or modified pastoral programmes, the greater the requirements for new curriculum development, collaborative planning and monitoring by teachers, and the re-assigning of academically-orientated time. Such requirements, as well as their substantial resource and in-service implications, may not be realistic and, here again, consideration should be given to ways in which 'personal development' aspects of teaching and learning within the 'academic' curriculum can be made more explicit and used as the basis for personal elements in recording systems.

Thirdly, the relationship between the various parts of schools' pastoral programmes is often unclear and it seems to us that this relationship needs to be clarified to enable the introduction of RoAs. We have little evidence to suggest that RoAs provided a coherent focus for these diverse activities, but such change may be more long-term.

Finally, our evidence suggested that some pilot schools were moving towards increased tutorial time and more structured tutorial work to enable or accommodate the personal element of recording. However, even in schools which had created the space on the timetable, and had given consideration to first principles and developed courses, materials and methodology, the task of integrating the recording element and the curricular programme had not been straightforward. Such integration clearly requires much more than the mere co-existence of the two elements. In our judgement further consideration still has to be given to whether, and how, these links can be strengthened if the expressed aim of grounding personal recording in the context of classroom work is to be realised.

At Rose School an ATW-type pastoral programme was established in the the Lower School before the advent of the RoA initiative and the pastoral component of the RoA was perceived to fit well with this. Thus personal recording was incorporated into the pastoral programme although additional time was provided during the pilot phase for one-to-one review sessions.

In the early stages of development there was little evidence that the existing pastoral programme changed as a result of this new addition. However, at the point when the school sought accreditation for a pastoral summative record for fifth years in 1988 it was recognised that personal recording needed to be better integrated with pastoral work — as an aspect of it rather than a bolt-on extra. Thus activities associated with developing the statement of personal qualities reflected the kind of exercises designed to enhance self-esteem that were a familiar part of Active Tutorial Work.

The issues which emerged in the relationship between recording objectives and the provision of learning opportunities in the context of schools' **academic programmes** were not dissimilar although they were complicated to some extent by the impossibility of isolating records of achievement from the many other influences being exerted upon academic departments at present. It was also impossible to distinguish between the causes and effects in an area as complex as curricular and pedagogical change, where new thinking in a particular subject area may well influence, and be influenced by, the principles and practices associated with RoAs. Furthermore, many of the changes in academic programmes had less to do with curricular provision in terms of coverage than with the organisation of teaching and learning, which we deal with in 2.4.3 below. Nevertheless, subject teachers in some case study schools began to consider how their courses might support a wider range of achievement than had previously been sought, for example, how they might specifically promote personal development in terms of personal and social skills (see 1.1.1 (b)).

Underlying this evidence of changes in the way that some pastoral and academic programmes were being conceptualised, we believed that we could detect the beginnings of movement towards the erosion of the traditional divide between pastoral and academic aspects of the curriculum. This was achieved less by structural changes in relation to the organisation of schools (although we also have evidence of this) than by a reconceptualisation of curricula along the lines described in the first section of our across-site analysis (1.1 **Coverage**). Thus some subject departments were beginning to accept the need to plan for personal aspects of achievement with academic curricula and, conversely, some tutorial staff were beginning to see their role as being concerned with oversight and support for the total educational experience of their tutees, including both academic and pastoral aspects. In our view this trend is to be encouraged because it holds the promise of a valid and coherent curricular basis for records of achievement systems.

At Poppy County High School it was proposed that a specified PSE period should be introduced in 1989 to support a redefinition of the roles of form staff and senior tutors as responsible for the oversight of the total development of pupils (academic as well as personal and social) but this was as much a result of a review of the whole curriculum as it was a response to the RoA initiative.

Within academic courses however there was some evidence that the requirement that subject staff should discuss and record comments of a personal nature stimulated them to consider the particular personal and social qualities that were, or might, be fostered within their specialist area. The need to find ways of coping with personal stress in relation to the demands of project work in Technical Graphics was a particular example.

2.4.2

Conception and use of tutorial time

As our last comment above indicates, we found a changing conception of the role of tutors of such interest that we would wish to endorse a general move to define the role of the form tutor (or equivalent) as having responsibility for a general overview of the totality of the educational experience of the pupil. This would seem to us to be the best way to facilitate the co-ordination of records into a composite RoA which gives a valid picture of the whole person. This is not to say, however, that tutors should not take a more particular interest in the personal development of pupils through routine pastoral contacts or through pastorally-orientated curricular programmes. However, if, as suggested above, subject curricula were to be reconceptualised to incorporate social and personal aspects in an explicit way, the function of pastoral programmes would need to change. Instead of being **the** area of the school's curriculum in which the personal element is dealt with, they would need to develop a supportive or service role in relation to personal and social aspects of educational experience elsewhere.

We cannot claim however, that our evidence provides us with many examples of such redefinition of curricula and roles. Many of our case study schools were a long way from reaching such a point in their thinking and a small number had yet to develop a conception of the role of the form teacher that went beyond the purely administrative. Thus for some schools any redefinition of the tutorial system would be a subtantial innovation with all sorts of implications.

Admittedly, for most case study schools, as far as the development of RoA systems was concerned, the major consideration had less to do with details of the role and function of tutors and more to do with the detailed organisational implications of incorporating tutor-pupil review sessions. Whether schools took a narrower view and defined the tasks of tutors in relation to personal experience and achievement only, or whether they took a wider view and associated the tutorial role with oversight of the whole of a pupil's educational experience, the need to find time for the reviewing process was regarded as crucial. Indeed, finding time to carry out all the new processes associated with recording was the most frequently cited obstacle to the success of the RoA initiative irrespective of the direction taken by the project in the school. Most teachers we worked with aspired to do justice to the potential of records of achievement in order to enhance their pupils' education as well as to satisfy their own professionalism, and as a result they felt frustrated and compromised when time constraints intervened. At a general level therefore we have an obligation to report that schools regarded time as the most important resource and many were not convinced that they could do the job properly without some enhancement in terms of staffing. Towards the end of the pilot phase the cumulative pressures on time associated with new initiatives such as RoA, GCSE, CPVE, and TVEI assessment and recording were so great that some schools felt they were reaching breaking point. **Time, then, was crucial**. However, beyond this general statement it is not easy for us to say how much time is needed or how it can best be made available. Amongst our case study schools there was considerable variation in the amount of time tutors, for example, felt they needed for review sessions which depended on how they defined their roles and the processes in which they were engaged.

One member of the pilot tutorial team at Columbine Middle School estimated that each of his interviews had taken up a full lesson, if he included the preparation beforehand and the writing up of notes afterwards. At this rate it could take a entire school year to work through a class of thirty children, with only one non-contact lesson per week given to tutors for extra work such as interviewing.

At Daisy High School some tutors found it difficult to limit their interviews with some pupils to the ten minutes originally allocated. The form tutor studied as part of the 'close focus' had little difficulty in interviewing two pupils in each 20-minute session, but other tutors reported in interview with the evaluator that they sometimes devoted the whole of a single 20-minute session to one pupil, using the interview for a slightly wider pastoral purpose than the profiling exercise alone, e.g. discussing problems at home, at school, with friends and so on.

We have already said that different patterns of project development carry different implications for the roles and responsibilities of staff in schools, for in-service training and resources. Analogously, the way in which the process of recording achievement is defined within the school will carry implications for the scale of the management task which ordinary form tutors expect to carry out, as well as for the resources needed to support this task. The more comprehensive and ambitious definitions demand greater commitment and expertise, which in turn require greater recognition and support.

2.4.3
Changes in the organization of teaching and learning

As we mentioned in passing in 2.4.1. above, the most notable changes which we came to associate with the introduction of RoAs in relation to academic programmes was a far greater tendency to organize learning into shorter-term units of work, with stated targets at the beginning and regular recording intervals on specified criteria.

These changes did not usually constitute 'modularisation' as the term is generally understood, since we have little evidence that pupils were able to choose from a menu of modules or that the modules were at some level interchangeable (i.e. no strict progression). Nevertheless, the organization of learning into units of work was for most schools a major modification to the academic programme. Such units were often of half a term's length, and each had its own organizing theme, as well as specified skills, concepts and knowledge which were accountable in the context of the unit. In this way the 'recordable' achievement, and the teaching objectives, for each unit of work, were expressed in the same terms and were therefore potentially more closely integrated. The approach also obviated to some extent the necessity of creating yet another assessment framework to accommodate RoAs.

These changes had substantial implications for both teaching methods and classroom management. We observed teachers taking more time to explain the objectives of units of work to pupils, pacing their work more slowly but frequently checking progress with pupils. In order to accommodate such changes, syllabuses were scrutinised to see whether content might be cut so that what was learned might be what was really essential and that it might be learned more thoroughly.

We also observed that in a number of schools where regular teacher-pupil discussion in classroom contexts was regarded as an important part of recording processes in subject areas the implications for classroom management were similiarly positive but far-reaching. For example greater emphasis was placed on individualised or group tasks, independent learning and/or team-teaching. These were regarded as important strategies not only in their own right but also as the only practical way to accommodate more frequent teacher-pupil discussion in lesson time, as discussed in 2.4.1.

At issue here is whether the introduction of new recording elements are dependent in the long-term upon major changes to the way in which curriculum and teaching is organized — for example, toward more unit-based courses and independent learning approaches in classrooms. If such changes are essential to enable pupils to develop and demonstrate the knowledge, skills and attitudes acquired in curriculum contexts, and to enable them to be fairly discussed and recorded in interactive processes, then serious consideration will have to be given to the substantial resource and in-service implications. On the other side of the equation, however, is evidence that the reorganization of curriculum and teaching can, in the long run, rationalise some of the current load carried by teachers and make potential savings in terms of time and energy because it encourages schools to go back to first principles rather than simply adding another task to the existing load.

Without substantial additional resources for development, but in particular in-service education geared to this sort of change, it would be difficult to insist that such changes should be integral to all RoA systems. Whilst we would wish to point to this as a direction worthy of development in all schools we recognise that some are not yet in a position to take such radical change on board even though they may be able to develop an adequate RoA system based on their existing arrangements. In our judgement, therefore, this is an area for advice rather than prescription.

2.4.4.
Need for whole school policies on curriculum, assessment and recording

As intimated in our discussion above, the introduction of records of achievement in some of our case study schools was accompanied by reviews of curriculum and assessment policies at whole school level. In some cases the RoA initiative stimulated general policy discussion in a direct way although other initiatives, such as GCSE, were also a powerful influence. The main thrust of this development, however, was to recognise and respond to a need to enshrine some of those aspects of curriculum and organization discussed earlier in a common policy to be agreed and implemented across the school. Our case studies provided us with a number of examples where the result of such discussions was an explicit whole school assessment policy (see 1.4.2 (a)).

Where such a policy was lacking we gained the impression that as, towards the end of the pilot phase, RoA systems in case study institutions were being firmed up, the need for such policies was increasing. Indeed we had evidence of considerable frustration in settings where decisions on policy issues to do with assessment were not forthcoming.

About half way through the pilot project at Nasturtium Tertiary College the co-ordinators realised that their work was seriously threatened because they were not having much impact on the Deans whose support was vital. It seemed to them at the time that, having appointed co-ordinators, senior management felt that they had met their commitment to RoA and assumed implementation would follow automatically.

This was not the case, and although the co-ordinators had a direct input into Academic Board where academic policy was agreed, the introduction of a college-wide RoA system required more than goodwill. In particular it needed to be taken on by the Deans as a dimension of their work with students for which they would be fully accountable. This they appeared to resist because they already felt overloaded.

As in many large organizations which aim to encourage wide representation and participation, the committee structures of Nasturium College were elaborate. The records of achievement project was monitored by a special Records of Achievement Sub-committee of the New Educational Opportunities Sub-committee of Academic Board. Evidently these committees were unable to come to any resolution on the issue of RoA management post-pilot, so a management sub-committee was set up just before Christmas 1986 by the Principal, Vice-principals and Deans, who also met together regularly. The first meeting was rather sticky but in the second, which the evalutor observed, the two co-ordinators were very clearly lobbying for an explicit management decision which would make an RoA system part of college policy on a given date and would make Deans directly responsible. On the last PRAISE data gathering visit they said they had at last achieved this end although not without a fair amount of blood-letting all round.

Given the multiplicity of assessment initiatives at the current time, we would expect something in the nature of an assessment policy to emerge in the near future in most of our case study institutions. We would regard this as an essential development although we recognise that the relative contributions to such a policy by school and local education authority will probably depend on local arrangements. In some of our case study schools it was clearly expected that the LEA would establish a broad framework in which individual schools would fill in the details according to their circumstances, whereas the system eleswhere was more devolved and schools could expect to formulate policy more or less for themselves.

2.5
In-service training

Throughout this report, we have repeatedly pointed out the implications, for in-service training (INSET) and staff development, of the evidence we have presented and of the issues we have raised. Clearly this aspect of the initiative is critical, and we should like in this section to bring forward some of those implications.

Our interest in INSET activities has been confined to their usefulness in helping us to complete the picture we have been trying to build up of RoA projects within schools. We have therefore been more concerned with **people's accounts** of their experience in INSET activities, than with documenting or evaluating the activities themselves. This has, in turn, led to the problem of how to interpret individuals' accounts of their experience. Inevitably, different people have valued different kinds of INSET, making it difficult to assess relative effectiveness and to identify generalisable issues for policy. The summary of project director's reports, which is found in Part Two of this report, also contains a section on INSET. This provides an overview of and commentary on various aspects of the training activities which have taken place within the schemes as a whole.

These difficulties are compounded by the absence of an acceptable definition of in-service training in the context of the RoA initiative. On the one hand, we should like to argue for a broad definition of INSET which encompasses not only those events specifically planned for training, such as county training days for prospective reviewing tutors, but also those occasions on which teams of teachers meet to plan, develop, evaluate and review the processes and materials comprising their projects. According to this definition the numerous working groups which have been convened within schools, including departmental and tutorial teams meeting to discuss their approach to recording achievement, can legitimately be described as training in the broad sense. We should also like to include the more routine 'training transactions' which take place almost casually between colleagues in an institution and between staff and outsiders. In particular, the informal discussions held between members of the central RoA team and individual or small groups of teachers within a school have possibly been the most frequent, but least conspicuous, form of INSET taking place. On the other hand, we would not wish to imply, by placing value on a wider range of activities, that **special** training activities could or should be dispensed with. All of these activities are potentially of value, and our main point in arguing for a broad definition is to reinforce the idea that all of them have a potential training function.

It is also important to distinguish between INSET where content and form are pre-specified by e.g. the LEA team, and INSET which is responsive to the needs of schools and teachers.

Generally speaking, we have documented INSET relating to general principles and philosophy, detailed development work, and skills training. Activities have moreover been provided at regional/consortium level, local/LEA level, and school level.

Regional activities have tended to concentrate more on dissemination of ideas and reports of progress between pilot schools than on training in specific techniques. These kinds of activities have also, because of their broad geographical scope, been attended by people as representatives of their schools or clusters of schools rather than by whole groups of people in schools.

Local activities have tended to concentrate upon training in specific techniques for whole groups of teachers, e.g. prospective reviewing tutors, and are held at some central point in the county and run by the LEA team. Interviewing skills have featured prominently in these kinds of activities, this area of RoA work being most often cited by staff as the one in which their own initial training and subsequent school experience prepared them least well.

School-based INSET covers a diversity of activities including all the informal training transactions mentioned above.

In many of our case study schools, teachers were involved in INSET activities at more than one of these levels, and sometimes all three.

One year into the pilot phase at Nasturtium Tertiary College the college team mounted a major three-day rolling programme of two-hour training sessions for staff and nominated students. Over the three days more than 100 tutors attended which accounted for almost 50 per cent of total staff. Each two hour session involved an introduction to the RoA project both at county and college level supported by video-recordings, active workshop sessions focusing on assessment and recording processes, and a concluding session encouraging staff to think about how they could best develop an RoA system in their areas. The co-ordinators emphasised their willingness to help with design but drew back from offering anything cut-and-dried, which some tutors would clearly have preferred. Other forms of in-service reflected this intention and co-ordinators spent much of their time working with groups of tutors on specific proposals.

Co-ordinators themselves were involved in in-service at county level and in the later stages of the pilot phase were deliberately involved by the scheme director in providing in-service within other institutions. This was a major strategy for dissemination within the county.

Most in-service training at Campion Upper School, was carried out in-school although the school co-ordinator and one or two tutors had attended co-ordinators' meetings and RPA related courses at county level, or visited other Suffolk schools or schools across the county boundaries. The tutor whose group was the focus of most of our observations had also elected to go on an independently organised counselling course because he felt that he needed more training in skills which he preceived to be required by the RPA process. The Head of Guidance pointed out a problem however in that there was no formal framework for the feedback of information from those who attended county or out-of-county courses. In other words the assumptions underpinning the 'cascade model', on which this pattern of INSET was based, did not appear to be fulfilled.

Within the school, in-service was focused on the year team meeting because issues were perceived to be different at different stages in the RPA process e.g. third year tutors worried about introducing the system, fourth year tutors about sustaining it, and fifth year tutors about statement writing. Although an hour or so, once in a while, was not considered adequate, tutors, especially probationers, were often reluctant to be taken away from their groups for longer; in-service could actually disrupt the relationships which were supposed to be the heart of the RPA process. School-based in-service meetings were mostly conducted by staff for themselves although the LEA team led the session on statement writing for fifth year tutors in 1987.

We are naturally reluctant to advocate a single model for the **delivery** of RoA-related in-service training, since delivery approaches will depend so crucially on local circumstances. Instead, we identify below what we feel are the critical training priorities emerging from our analyses elsewhere in this across-site report.

(a) Arising from our discussion in 1:1 of the complex notions of achievement which underline RoA systems, and of the far-reaching curriculum implications inherent in different conceptions of the knowledge, processes, skills, attitudes and experience covered by RoA systems, there is a clear need to provide teachers with opportunities to **analyse the curriculum** in a fundamental way, so that their recording systems can become grounded in valid conceptions, and so that there is a conceptual and practical link between teachers' notions of the curriculum, the opportunities they provide in the classroom for learning and development, and the terms in which achievement is recorded.

(b) The structured subject-based assessment frameworks which have been devised as part of certain RoA systems have, in their provision of elaborate networks of criteria, stages and levels, presented a daunting prospect to many teachers, as we discussed in 1.2. It is crucial that teachers become fully aware of the underlying rationale of such **assessment frameworks** and of the practical operation of them in the classroom.

(c) Allied to this is the need to provide opportunities for teachers to think through and improve their techniques for **training pupils in the 'art' of self-assessment** (1.2). This is particularly important if the enormous educational potential inherent in this process is to be realised.

(d) Arising from our discussions in 1.2 and 1.3, of the central place which statement-writing has assumed within most RoA systems, there is a clear need to provide teachers with opportunities to acquire more sophisticated techniques for **summarising** often diverse forms of information about a pupil. The growing emphasis upon verifiable portrayals of pupils supported with evidence places new and additional pressure on teachers to handle evidence judiciously in order to represent pupils fairly, succinctly, positively and in a way which will be immediately understood by a non-educational audience. Allied to this is the need to prepare teachers for the task of **training their pupils** in this crucial area of personal summarising and the handling of evidence.

(e) It has almost become a truism to say that there is a need to provide teachers with opportunities to develop **one-to-one discussion skills**. Whilst we accept that teachers are not unaccustomed to talking to individual pupils, there are a number of special features of the characteristic RoA-related interview which make it quite unlike the circumstances in which teachers normally converse with their pupils. We identified several of these special features in 1.2 and 1.5. We would single out for particular mention the importance of **context** (1.2), and the social and cultural implications of discussing pupils' activities, interests and experiences outside school (1.5).

(f) There are also a number of training implications arising from our discussion of management issues. Perhaps most important is the need to prepare senior managers for the delicate and complex task of **developing whole-school policies** which incorporate RoAs and the myriad of other related initiatives currently being introduced (perhaps especially GCSE).

PART TWO

Across-scheme analysis of reports from project directors and local evaluators

by Patricia Broadfoot and Desmond Nuttall

Contents

This part of the report summarises the main conclusions of the reports written by the project directors at the conclusion of the three years of development work supported by Education Support Grant. In some cases the project directors also acted as evaluators of their scheme; in others, independent local evaluators were appointed. Few of the local evaluators had completed their final reports by March 1988 (the date by which the project directors had to submit their final reports), but interim reports from some local evaluations had appeared at various points during the previous year. The findings of these reports have also contributed to this summary, though as often as not they had also been drawn upon by the project directors in their final reports, and were appendices to them.

The amount and nature of the evidence to substantiate the conclusions reached by the project directors were therefore variable. As a consequence it proved impossible for us to **evaluate** the reports and to reach judgements on the factors and particular approaches that facilitate the development of good practice in recording achievement.

The difficulty of evaluating the reports was exacerbated by two factors. The first factor was the difference in interpretation of the notion of a **pilot** scheme. No doubt partly as a result of the teachers' action during much of the first 18 months of the development phase, few of the projects kept to their original timetables and plans. Furthermore, and more importantly, the projects did not view themselves primarily as part of a venture whose main aim was to gather experience and evidence on which national guidelines could be based. As a result, evidence on some issues, such as the factors that contribute to the credibility of the summary record, is very thin.

The second factor was the inevitable consequence of the preparation of reports by many different writers. The reports from the pilot schemes were based on guidance from RANSC which requested the project directors to provide evidence about certain issues (e.g. resource implications, IT, INSET, user reaction, and implications of gender and culture differences) and any others they considered to be important, and then to provide a retrospective evaluation of the development work which highlighted the lessons learned. This guidance resulted in some useful comparability despite marked differences in the style of the reports, but inevitably project directors were selective in the detail of the issues they addressed, and evidence of good practice from one scheme did not necessarily have counterparts reported in another to illustrate the range of possibilities. Furthermore, the reports of the local evaluations conformed to no such standard guidance. While this was on the one hand a considerable strength, since it allowed unforeseen issues to be raised and explored, it did make any attempt at comparison between schemes much more problematic. In preparing this part of the report, we have therefore been very conscious of the variable nature of the data used and the need to exercise extreme caution in placing equal weight on reports whose different provenance necessarily reflects considerable variation in the objectivity that they can exercise.

As a consequence, no attempt has been made to pass judgement on particular schemes, or to compare and contrast their effectiveness. All that this part of the report attempts to do is to summarise and analyse the experience of the schemes, as seen through the eyes of the project directors, assisted by their local evaluators in some cases. Many common themes emerged and have been classified under virtually the same headings as were used in the corresponding part of our Interim Report. The only notable addition is that of a section on equality of opportunity (Section 3), an issue on which evidence was particularly sought. The issues are categorised under similar headings to those of Part One, and are presented in a similar order, but they are not identical. Because of the perspective of the project directors, they give more prominence to issues of management and issues that go beyond the school. There is, of course, overlap between issues raised under many of the headings. For example, Section 8 (**The Role of the LEA**) concentrates on the ways in which the LEA can promote co-ordination of different initiatives and gives less space to the ways in which the LEA can support the development of RoA, since this support is covered implicitly or explicitly in other sections such as 4 (**School Management Issues**), 6 (**Information Technology**) and 7 (**INSET**).

References to specific reports from pilot schemes have been included to give some indication of the range of source of material upon which we have drawn in the analysis. This referencing has sometimes been done by name but usually by number according to the following:

Dorset
— Final report to RANSC (1)
— Second interim report of the local evaluation (2)

EMRAP
— Final report to RANSC (3)
— Northants preliminary evaluation documents (4)

Essex
— Final report to RANSC (5)
— Evaluation report and notes (6)
— The Achievements of Robert Arthur Essex (7)

ILEA
— Final report to RANSC (8)
— Evaluation of the first London Record of Achievement portfolios — 1987 (RS 1184/88) (9)

Lancs
— Final report to RANSC (10)

OCEA
— Consortium final report to RANSC (11)
— Coventry final report (12)
— Leicestershire final report and evaluation material (13)
— Oxfordshire final report (14)
— Somerset final report (15)

Suffolk
— Final report to RANSC (16)

Wigan
— Final report to RANSC (17)
— Emerging issues — A discussion paper (18)

WJEC
— Final report to RANSC (19)

In several cases, the final reports included local evaluation reports among their appendices. Many of the reports were very extensive: for example, the OCEA reports (numbers 11 to 15) and their appendices consisted of 1420 pages, not sequentially numbered. (As a consequence, we have not given page references of the direct quotations from reports.)

Where no reference is made to a specific report or reports, it should be understood that the aim has been to draw out the consensus view emerging from pilot schemes – where such a view was discernible in their reports – rather than to introduce our own conclusions and recommendations.

1 Content, coverage and principles of records of achievement

Most of the schemes have explicitly stated principles to guide the process of reviewing and the production of records of achievement. For example, there is a common requirement for regular discussion between pupil and teacher, which is seen as the crucial ingredient in the formative process. There is less agreement over the age range of the pupils to be involved: for example, EMRAP required schools to involve all children of secondary age, whereas the Suffolk scheme limited the process to years 4 and 5 in the secondary school. The WJEC scheme preferred schools to involve all pupils but insisted only that those in years 4 and 5 participate; the WJEC also specifically required the inclusion of pupils with special educational needs. In respect of the summary record, there is agreement that the pupil should own the document and control its use.

However, the criteria were not spelt out in the same detail by each of the projects. For example, some projects such as EMRAP and WJEC made it clear that it was appropriate to include negative comments in the formative process but not in the summative records, but only WJEC specified who should make the comments (the teacher). This points up the more general issue of the degree of specificity and detail in criteria which are needed to promote good practice without infringing local autonomy.

The terminology used across projects varied and was not explicit in all the reports. Terms used include: 'personal achievements, experiences and targets' (3); 'experiences of work', 'experiences of school', 'out-of-school interests and activities' (16); 'personal interests and achievements' (19). Other examples are 'work experience', 'achievements outside and inside school', 'achievement curricular and extra-curricular', 'experiences and achievement', 'experiences'. Here we look at three major categories: academic attainment, cross-curricular skills, and personal skills and qualities, which represent the coverage of most schemes (but which would seldom be used as titles on records).

All of the schemes except Suffolk appear to have developed formative assessment in this area. in some schemes these were based on comment banks (10, 19), whereas other schemes explicitly rejected this idea (8, 17). ILEA for example suggested:

'The conceptualisation of attainment in pre-specified forms, unless based on extensive analysis of actual performances, appears however to present major difficulties . . . The creation of statements that have sufficient detail to make them more meaningful can lead to an assessment scheme in which the sum of the parts no longer seems to represent the whole, and which is cumbersome to manage and difficult to communicate other than to the students and teachers involved . . . At issue is how learning aims and objectives are shared with students and what we mean by aims and objectives. The setting of curricular aims and objectives is generally approached in one of two ways. The first of these is to pre-specify precisely the learning outcomes of each course and state what skills the student will be able to demonstrate at the end of it. This is perceived as a rational approach to curriculum planning which enables teachers to make it clear to students what the intentions of their learning are. The course objectives match the asessment criteria exactly. The second approach seeks to create a framework for learning by establishing the areas of experience and the processes of learning to which the students will be introduced. This approach recognises that students have different starting points and that their intellectual development relates to the context in which they understand the ideas and experiences with which they come into contact at school. The opportunity for individual students to proceed creatively is seen as part of a process of educational development which aims to expand the individual world of meaning in which each student perceives and understands. The assessment attempts to describe the complexity of individual student's responses to the curriculum which is being presented.' (8)

Records of Achievement

The growth of formative recording has led to a questionning of the need to continue with traditional methods such as marks, effort and merit scores etc, and even the traditional role of school reports and parents' evenings. The formative process contains several elements: classroom-based discussion involving clarification of learning expectations, review and reflection from teacher and student, and planning the next and future steps.

'This in turn leads into a more formal review and reflection stage of a similar kind. The interpretation and performance in relation to national attainment targets ought to be complementary to such an approach.' (1)

Student self-assessment was mentioned by several of the schemes as a way of encouraging pupils to take more responsibility for their own learning and as a way of providing a starting point for one-to-one discussions. However, students appear to be influenced by expectations they perceive their teachers to have of them and respond accordingly; they also tend to use norm-referenced judgements and to be reticent about their successes perhaps because self-praise is frowned on in our culture (8, 18). Certainly it can be a useful exercise for teachers to examine the relationship between what they have set out to teach, what they have assessed and the student's perception of what the learning is about. Group self-assessment has also been tried: some classes have made videos of themselves and then have discussed how they worked, what they achieved and what they will do better in the next project (8).

1.2.2
Cross-curricular skills

Many projects have attempted to promote the assessment of cross-curricular skills, which have been defined as 'skills and concepts which are common to a number of discrete subject areas' or 'those skills and qualities which are not specific to any one area, such as inter-personal skills, language development, learning skills, etc.' (17). Various problems have been met. One is the daunting length of the lists of such skills that have sometimes been developed — a problem solved in one case by limiting the list to a core and allowing each subject to add more specific skills (8). Another is the varying interpretation of the same term in different contexts, and the consequent variation in the evidence and criteria employed, leading to difficulties in explaining the concepts to students. Collating and aggregating assessments of nominally the same skills from different subjects and contexts has also presented problems in some schemes. It seems fair to conclude that no project has yet devised a wholly satisfactory scheme for the assessment of cross-curricular skills.

1.2.3
Personal skills and qualities

Issues in this area are many: for example, who should assess pupils' personal qualities — the teacher alone, the pupil alone, or some negotiated assessment? Should a checklist of headings be used and at what level? The range of practice in this respect stretches from the provision of pre-specified checklists for teachers to use (19), through prompt lists and aide-memoires (3) to no headings at all (8). A further question is the form in which the recording should take place (and whether grades or a comment bank approach should be used), and the extent to which evidence and context is provided. Furthermore, what opportunities were given for such qualities to be displayed? Underneath these questions is the more profound question of exactly what definition of 'quality', 'skill' or 'achievement' is being used. The Wigan report pointed to the need to distinguish between 'personal qualities' and 'personal achievements' and questioned the legitimacy of recording personal qualities without making explicit the value base upon which that process is founded. It suggested that personal qualities are best recorded within the context of the learning situation, and that some qualities such as 'honesty' should be proscribed, as should recording without the provision of a substantiating evidence and the opportunities for all young people to develop such qualities.

The concerns in this field are further exacerbated by the differences in the nature of students' out-of-school activities, some of which may be 'low status' and some which may be 'high status', leading to the possibility of further disadvantaging those already disadvantaged by socio-economic status. This issue becomes even more sensitive in moving from the formative to the summative (3).

There is clearly an enormous variety in formative recording from school to school and often from department to department within a school. The only general point it appears possible to make is that there is an unresolved debate over the merits and implications of statements versus lists of specified criteria, and over the associated issue of the importance of context and evidence.

1.3
The formative record and the formative process

The same variation is associated with reviewing practice in the schemes, but it is clear that all schools are making some provision for review, at least in the pastoral domain. The frequency, duration, location, and organization of these activities cover the complete spectrum of activity from recording and discussion in individual subjects, recording and discussion in personal reviews with a form tutor, group discussion in personal reviews with a form tutor, group discussions with a form tutor, to personal recording without formally arranged discussions. There would appear to be some tendency for more time to be given to form tutors to make this possible, but seldom to subject teachers.

1.4
The summary record

The preparation of the summary record and the constraints surrounding its length and coverage present problems in many schemes. For example, there is uncertainty about the best way to record curricular achievements on the summative statement with particular regard to the language and length of such statements (11). An attempt to reconcile the two purposes of formative and summative reporting resulted in many members of staff feeling unhappy about the way in which profiling processes had become dominated by the reporting system (10).

The contents and coverage of summary documents are illustrated by the following selection:

'Examination results, new forms of academic attainment, personal qualities, evidence of pupil attainment gathered from outside the curriculum.' (17)

'Course details, course achievements, personal qualities and skills, pupil's statement and extra curricular and other valued experience.' (3)

'Courses of study, attainment, personal strengths, experience and interests, exam entries and attendance.' (16)

'List of subjects and level, cross-curricular subject skills, personal and social skills and qualities, interests and achievements in and out of school, work experience report, exams and other certificates.' (19)

'Information about the school and the record, a description of the content of a portfolio and a list of validation board members. Subject-specific statements including pieces of work, photographs of events and objects, work experience record. A summary statement constructed by the tutor and student and a selection of certificated achievements including exam certificates, graded test certificates, sporting certificates and community activity certificates.' (1)

'A student statement, which describes a student's achievements inside and outside school and reflects positively on their personal and social skills, their interests and their aspirations. A school statement in which teachers give a positive picture of a student's achievements. Samples or photographs of good pieces of work. Certificates gained in competitive and non-competitive examinations and tests, inside and outside the school, summary subject profiles, unit credits, graded assessments and public examination certificates (8). (Not all these will appear; in the first year that Records were awarded, graded assessments were included in 3 per cent of the portfolios and unit credits in 14 per cent. In non-academic areas, 28 per cent of portfolios included some form of certificate with the most common being sporting results.' (9)

In this last case, the school statements were not negotiated between the student and teacher in many cases (9). In most of the other reports little mention was made of negotiation, and the notion appears to be rather taken for granted. For example, in one EMRAP county the final report simply said that the record is negotiated. In another EMRAP county, the pupil selected personally valued achievements, and in Suffolk the pupil assembled the information which was then discussed with the teacher prior to writing.

The length of the records normally ranged from four sheets of A4 to eight sheets of A4. Sometimes there was no prescription of length but rather a detailed description of the principles underpinning the production of the record and the use of common water-marked paper and the LEA crest to ensure the authority of the record (17).

There was a general emphasis on the need for evidence to accompany the summative record in some form to ensure comprehensibility and validity for the user. The timing of the preparation of the summative reports was also referred to in a number of schemes. The need for some kind of interim summary record to be available in the autumn term of the fifth year which can then be updated was stressed by many, and the utility of word processors to allow for regular updating was mentioned. All schemes took the view that the pupils must own the summative record. The life span of the record was explicitly regarded as short in some cases (e.g. 3, 16).

The major area for debate seems to be the principle that negative statements should not appear in summary records. The WJEC noted that there were initial worries about this apparent dishonesty, which disappeared when teachers understood the philosophy better though they were still worried as to whether users will be so easily convinced by the arguments. This is discussed further in Section 9 (**User Response**). A related problem is the contradiction between the positive, collaborative approach to assessment fostered by RoA and the realities of competition in public examinations (8).

Most of the pilot scheme reports suggested that there has been a favourable response from pupils and some improvement in motivation, though this is qualified in a number of different ways. The WJEC pupil survey, for example, recorded a generally positive response to RoA, pupils being both 'Pleased — 91 per cent' and thinking it 'Fair — 80 per cent'. Whether or not RoA have improved general motivation was less clear however; evidence from the WJEC teacher survey and interviews with heads/co-ordinators suggested that among the teachers, 26 per cent thought 'Not all', 54 per cent 'A little' and 15 per cent 'Quite a lot'. Improvement in motivation was particularly likely to be the case if pupils were involved in the assessment. As regards the review session itself, teachers in the WJEC scheme reported that 9 per cent of pupils showed 'enthusiasm', 77 per cent 'willingness', 14 per cent 'reluctance' and none 'hostility'. The Suffolk evaluation data suggested that pupils found reviews helpful for 'setting targets' and 're-thinking about oneself'. Certainly, the experiece of Wigan sums up that reported by many of the schemes:

'Motivation in the classroom appears to be dependent on how much responsibility is given to the pupils. In subject areas where they are encouraged to take responsibility for their learning and to be involved in reviewing and assessment, they are very highly motivated. Where the subject is teacher-controlled there tends to be little change. However, some schools have observed that the pupils are becoming more confident and 'able to challenge' often in a very mature manner, often in those areas that are more traditional. This is directly attributed to the recording of achievement.'

Evidence of improved motivation includes: improved attendance in the fifth year, fewer cases of vandalism and discipline problems, and improved achievement across the board (17).

There were also indications that helping to make pupils aware of their learning targets and involving them in self-assessment, together with dialogue between student and teachers as part of the assessment process, lead to greater student self-esteem and to higher levels of expectation on the part of both students and teachers (5).

Students responded well to having more say in their learning from deciding how the work is to be approached through to reflecting on it afterwards, but this greater control had to be genuinely given to them rather than being a cosmetic exercise (13). The same report identified the impact in the classroom as:

1. Pupils more actively thinking through approaches to problems and trying them out.

2. Working in groups and discussing with other students what they are doing.

3. Looking less to the teacher for answers in the initial stages of a problem-solving exercise, and

4. Beginning to get used to the languages of assessment and review and actually assessing achievements and progress (13).

Many schemes referred to the important contribution of language to the clarification of the student's role. They need to be given the vocabulary to describe what they 'know, understand and can do' as well as 'needing to feel national value for the work involved' (17). The Lancashire report took this point further in stressing that:

'until teachers have reached a degree of clarity and confidence in their ability to articulate students' achievement they are not in a very strong position to help students in their own struggle to reflect on and define their own progress and learning. This is because they are not in a prepared position to be able to pass on language that a student can use within their own thought processes. What use is it to say to oneself that, 'I am a B− or D+' or 'That piece of work was 7/10'?

To summarise, it is clear that pupil response varies directly with the degree of organic change in teaching approach associated with records of achievement, that there is value in students understanding their achievements better, being more concerned about their progress, feeling a sense of achievement about what has been learned and understanding that learning better, and that all this depends on the teacher's ability to articulate the processes involved and to provide opportunities for genuine sharing on the part of pupils.

2.2 Problems

Several of the reports indicated that many students, regardless of ability, sex and ethnic origin, were demotivated by formative recording that involved extensive writing, which was seen as extra work. As a result, many of the schemes have moved towards a range of non-written ways of recording including discussion, graphics and displays (8, 12, 13). This is perhaps one aspect of a more general problem of maintaining the commitment to personal recording and logbook-style records. OCEA/Leicestershire in particular reported the difficulties of keeping pupils motivated with continuous recording and found equally good summative statements where a much less extensive personal recording provision had been made (13). Lancashire also reported that some students were bored by the completion of student self-assessment sheets — mainly when they were of similar format and similar questions were asked, such as: 'What have you learned?' and 'What have you enjoyed?' (10).

Another issue arises where achievement is perceived and defined within a middle-class value system. In such circumstances, those students who have traditionally not achieved will remain alienated from the education system and RoA will be irrelevant to them.

This will be exacerbated by the apparent lack of substance to the achievements recorded within the summative document . . . INSET is needed to help teachers be aware of this problem and to help to overcome it.' (17)

It has also been reported that some students felt themselves to be under constant scrutiny, with their every move being monitored (17). Schools and teachers need to be careful about the amount of recording they attempt so that it is not intrusive.

Students often seemed to enjoy recording, especially where they could see a link to the preparation of the summative document and thus employment (5). However, problems arose where teachers did not themselves understand the purpose and potential of the process. This led to less value being put upon it by students and by their parents. Several reports emphasised the importance of induction for pupils to help them understand the procedures involved and in particular to explain the meaning of criteria and lists of personal qualities if pupils were to use them with understanding (e.g. 3, 19).

Reviewing also had a mixed response. In the WJEC survey some pupils reported an initial nervousness, which could be removed by starting interviews early in their school career, by just letting pupils talk, by using small groups for reviews, and by stressing that information would not be used against them. There was some evidence that pupils were unsure of what personal qualities were being recorded and were unhappy about not being able to discuss them with the teacher. Some pupils said that they did not know if they were being assessed and usually found out from the form tutor whereas they would have liked the subject tutor to tell them (19).

In one evaluation study of a college, staff reported that some students found the processes of profiling⁵ and reviewing threatening and bewildering. They commented that students had been overwhelmed by the one-to-one review and diffident about negotiating comments, responses which they interpreted as the product of student immaturity or passivity. Some found that conducting the review with students had been tedious and unproductive, "like getting blood out of a stone." Similarly, "It is not easy to get students to comment fruitfully, they always tend to say what will please or will not offend the tutor." (10)

As that study pointed out, the crucial issue is that of 'power' in the negotiation process. The delicacy of the balance is illustrated by that report which described one way in which reviewing was conducted, with the teacher at the review explaining his feelings about the student's progress and then examining the student's claims of core competencies. The teacher then commented on the student's perceptions and asked the student for comments about the views of the teacher. This led to the negotiated statement. The teacher wrote it but tried to make the wording the student's. In addition to the issues of 'power' and 'control' implicit in this description is the issue of the variation between the way in which different numbers of staff tackled the issue of negotiation. Students found themselves "empowered in some review sessions and subjected to critical assertion in others". This had caused conflict for them.

One final point about reviewing concerns the hazard inherent in the practice of one-to-one dialogue in private (17). It is suggested that LEAs may need to address the question of a 'code of practice' for such discussions to safeguard all concerned.

2.3
The summary document

It is easier to refer explicitly to pupil response to the summative statement. The Essex report gave a full description of this as follows:

Many pupils who received an Essex record of achievement last summer were very generous with the praise for the documents which they received. All remarked on the quality of the document which was described by many as being most impressive. Many students remarked that it showed them in a very good light and that the teachers had clearly tried to support them. One student remarked, "It's nice that only positive achievements are included, otherwise it would have been much bigger." Many students remarked on the role of personal recording and stated that they wished, in retrospect, that they had devoted more time to building up a record . . . By contrast it is too early to ascertain the views of the current fifth year with respect to the finished product. At present their views are mixed, with some apparently begrudging the time it is taking to produce personal statements and assist with the record of experiences, others showing enthusiasm.' (5, p.15)

Essex pupils were, however, clear that for this to be the case a summative document must contain no negative information, although they were not against the school issuing confidential references alongside the summary document. They would like to inspect the latter for the sake of accuracy and have the right to correct information which was incorrect. However, unlike the pupils who left school, potential A level students who stayed at school regarded them as better than reports but largely an irrelevance to their future needs. These same A level students and those on equivalent courses were 'critical of recording outside their course under the category of personal experiences and strengths, arguing that experiences outside the college were private and had no place in the recording procedure for RoA.' (5)

Other schemes reported a generally favourable response to the summary document. The WJEC survey found that 78 per cent of pupils liked the summary document the way it is. Suffolk pupils felt that it would help with interviews for jobs and entry to FE, that it was good to look back upon, allowing the pupils to get a clearer sense of their own identity and to say things in the way they wanted to. OCEA/Coventry reported that:

'it gave new impetus to recording, almost it seemed because students realised the possibilities it represented (being told of the potential beforehand by teachers had a good deal less impact than actually seeing the completed summary), which should not be surprising about learning by experience.' (12)

In Wigan, some high ability students who were disaffected before have been won over by the prestigious impact of the summative document (17).

One qualification to the generally positive impact of producing the summary document is the need to educate teachers to recognise success and achievement (12). Schools that have developed ways of publicly recognising success, such as photocopying the best piece of work to be kept in a portfolio, reported that the involvement in this process of teachers, students, friends and parents had an influence on the way teachers view success.

Taking this point even further, the Wigan report indicated that the traditional structure and management of many of their establishments:

'militated against their recording achievement philosophy and feel that the move away from setting, streaming and labelling indicated their valuing of young people and was a move towards treating them more as individuals. In the words of one deputy head, "The situation is one of glasnost in that RoA is allowing an opening up where staff can explore educational issues".' (17)

Thus there would appear to be two basic elements in pupil motivation, the first being the extrinsic motivation of having an impressive and apparently useful product which the young person can take away from the school and which will help them in the world of college and employment. The other, intrinsic, element comes when teachers can sensitively change their approach to classroom practice and to reviewing progress with pupils, to provide a process which recognises the integrity and quality of pupils in these activities. When, however, the school as a whole, in its ethos and strategies, or individual teachers and departments fall short of this ideal by presenting pupils with opportunities which are still teacher-dominated in terms of assessment and reviewing or by presenting pupils with procedures they cannot control and do not understand, then the impact of RoA processes is likely to be negligible and possibly counter-productive.

Evidence is slowly beginning to emerge of substantial differences between pupils of different sex, race and prior attainments in their approach and attitudes towards records of achievement.

Wigan reported that girls were regularly underestimating their achievements when assessing themselves, less so boys, who were more inclined to over-assess their achievements. An under-expectation of achievement by students may also be reflected in the targets that they set themselves for future work in discussion with teachers (17). ILEA has also found that girls are more likely to allow their abilities and achievements to go unrecognised and even to deprecate them: "for girls in some cultural groups it is seen as a negative attribute to talk about one's strengths and achievements" (8).

In one ILEA school, monitoring by the staff found examples of sexism occurring in the social and personal categories. The school has sought to overcome this by developing 'general', 'social' and 'personal' statements. Another school monitored the school statements before the portfolio was collated, particularly for stereotyping by race and sex.

By contrast, the ILEA experience supports other research findings (notably those summarised in the evidence given by the Equal Opportunities Commission to the DES Task Group on Assessment and Testing) which suggest that boys do better than girls in multiple-choice questions but not in free response or essay questions. Girls, by contrast, usually respond well to the opportunity to write discursively about their work and enjoy discussing their profiles with their peers and their teachers (8).

'Boys are often less conscientious about recording, girls produce a higher quality presentation. Boys tend to be more confident in recognising their worth while girls feel more confident than boys in personal reflection. Generally, boys appear to think of an achievement quite quickly and sometimes have difficulty in deciding which to select. In two sessions that were observed, the girls struggled in the beginning to think of a single example and the facilitator had to work with them as a separate group. What emerged was a vast range of achievements, experiences and talents which they had not recognised for themselves. Because of these worries, in one school a 'Well Girls Group' has been set up and is creating a need for INSET on gender for staff.' (17).

In ILEA too, some teachers of boys reported that they sometimes appear to experience difficulty in expressing themselves freely and to welcome the support of prompt sheets. An evaluation of the first London Record of Achievement portfolios produced in 1987 showed that girls were significantly more successful than boys at compiling examples of good work and certificates and other evidence of their achievements. The evaluation also showed that girls were significantly more likely to include details of charity work and family responsibilities than boys (8, 9).

'It also showed that in those cases where teachers predicted the future career potential of students, 70 per cent of the comments related to boys and only 30 per cent to girls. It is possible for girls to experience positively intended stereotyping when they are praised for being friendly, helpful and compliant. When comments are made on the social and personal skills that support learning, there appears to be a tendency to focus on those that reinforce a passive role for girls such as their ability to listen and on the neatness and presentation of their work. Limited expectations of students can be revealed in this way, through the language that is used to describe their achievements.' (8).

Evidence from Wigan suggested that boys have more time spent on them while girls need just as much and, furthermore, that boys were challenging more and actually modifying statements as a result of dialogue. Wigan felt that girls need training to be assertive. Some strategies were being taken to counteract these dangers; for example, one Wigan school changed its class registers to alphabetical order, while in another the first RoA work booklet was changed after a group of staff pointed out the gender imbalance and stereotyping in the illustrations.

3.2
Race and language

Relatively little mention was made of race differences in the project reports, but some reference is made to language. ILEA in particular has taken a good deal of trouble to provide materials and opportunities for students to work in their first language, e.g. some of the materials for the London Record of Achievement, such as the booklet explaining it and the booklet to help students prepare their statements, have been translated into 11 languages. However, only one statement in the sample of 259 issued in 1987 was written in a language other than English in ILEA (9). It also appears that bilingual students appear to benefit more from discussions in pairs, because of the opportunity to talk more and to use two languages if the other person is also bilingual and speaks the same languages to some degree. Experience has shown that, for students and their families whose first language is not English, preparing the student statement in their mother tongue has been a significant and positive experience. However, there appears to be some concern that pupils who admit their first language is not a European language will be disadvantaged rather than have their mastery of two or more languages recognised as an achievement, although this is denied by employers (8).

3.3
Class

Considerations of social class had even less attention in the reports than those of race and sex. There is, nevertheless, a strong indication that, where achievement is perceived and defined within a middle-class perspective, this will be alienating for many students. The ILEA experience of using RoA in off-site support units is significant in this respect since, although there would appear to be a number of structural advantages such as the high teacher/pupil ratio leading to more time to work with individual students, the difficulty is that many students have little to show when they arrive at the unit, perhaps already in the fifth year, and do not have the time, or in many cases the will, to make the necessary progress to produce an impressive RoA. Indeed, many students are still deeply hostile to the educational process and this cannot be overcome simply by providing RoA experiences. Furthermore, to show how much progress has been made it is sometimes necessary to spell out problems that there have been in the past. Many students are also severely disaffected and have behaviour problems, which makes it very difficult for staff to work closely with them. In this situation, students working with each other seems to have many advantages. Other schemes reported that for severely disaffected pupils the RoA makes little impact.

For low achievers in particular, as revealed by the experience of many of the special schools involved in RoA, there is evidence of significant benefits to be gained from profiling (1). In Wigan some pupils who lacked writing skills were encouraged to experiment with different types of recording, tapes, photographs, word processors etc. Not using a system of headings in summative reports can help to avoid gaps that might look pejorative. Non-written forms of recording were also used for these reasons in other schemes, and the use of word processors has also been particularly helpful.

It is to be expected that evidence will be slow to accrue in this particular area. As with pupil motivation, it is a much longer-term and finely-detailed issue which may not be immediately apparent. However, it is worrying that not all reports even addressed the issue of pupil differences, desite the invitation to do so. It is quite wrong to assume the RoA will have the same impact on pupils with very different characteristics, abilities, interests and attitudes.

There is a general agreement that the school co-ordinator is a vital ingredient in the successful introduction of RoA and that he or she must be a senior member of staff, a deputy head or member of the senior management team (e.g. 3, 10, 19). The WJEC further spelt out the importance of the co-ordinator's commitment and enthusiasm, of having time to do the job, and of having secretarial assistance for it. He/she must be a good organizer, able to run INSET and have an office with computer support. In short, the co-ordinator 'must be clearly empowered to fulfil the designated role' (11).

The list drawn up by Wigan school co-ordinators also reflects the kinds of personal qualities that such a person requires.

1. The ability to consider and appreciate the view of others and to recognise everyone's part in the system, including the pupils'.

2. Management skills, including the ability to motivate and support colleagues and to persuade them to consider all points of view.

3. Sympathy and empathy in recognising where teachers are in their professional development.

It is clear that the commitment of the head and the senior management team is also central to successful innovation (e.g. 1, 3, 16). Dorset identifies two levels at which such support must be provided: commitment to and support for the underlying philosophy of RoA and, secondly, recognising in a positive manner the practical needs arising out of development work, such as releasing staff for in-service training and meetings. The commitment of the senior management team can be recognised as a reflection of a fact that the RoA has been fully integrated into the ethos of the school or college so that it is not a discernible entity in itself.

Another report put it thus:

'Records of Achievement must be integrated into curriculum planning along with assessment, recording and reporting, as well as the pastoral system. This requires long-term planning and structured strategies. Without the commitment and understanding of the senior management team, RoA will never be able to stand on their own, never mind being an integral part of the system.' (10),

There is considerable indication given in the reports that successful schools have been moving more and more towards a whole-school approach. Whereas many schemes started with a single cohort of pupils, pilot phase experience has shown how important it is for all staff to understand the issues and implications of RoA in order to bring about the necessary changes in schools as a whole. Only where it is perceived as a whole-school development have RoA appeared to have fully lived up to its intentions (8, 11).

Part of the whole-school development issue is that of ownership. The OCEA/ Leicestershire report illustrated this in their example of how distressed staff were when the criteria for one of the G components were changed, with little prior notice, at a point when they were just beginning to feel more confident and in control.

'Student reviewing and recording did not appear to raise the same anxiety, perhaps schools had largely to 'invent' it for themselves. Beyond the fairly stringent principles, there are no strictures on design or implementation of student reviewing and recording. In designing from scratch the schools had ownership from the start and therefore felt in control. This is not to suggest that there were no problems with student reviewing and recording, in fact there were many, but schools felt better equipped to handle them.' (13)

The need for development to be primarily 'bottom-up' is stressed by many of the schemes (3, 8,16,19). Both Suffolk and parts of Wales had had experience of existing systems and their reports admitted that this made it difficult to get teachers to change. WJEC reported that even when a package already existed, many teachers preferred to develop their own system, even though this wasted time. This raises the whole question of how far schools can learn from others and how far they need to go through the developmental processes themselves. But there is no doubt that integrating RoA with the larger work of the school and into a successful management of change strategy is seen as essential. Institutions that are particularly diverse, such as colleges or split-site schools, or very large have found particular difficulty in thoroughly familiarising those who have to own and operate the scheme. One solution to this is an RoA working party. In one ILEA school, for example, all seconds in charge of departments "are responsible for LRA liaison and they meet regularly as a group with a member of senior management in the Chair. This approach has proved particularly effective in ensuring wide dissemination and ownership." (8)

Wigan offers a number of strategies:

'An organic whole-school approach is facilitated by providing the opportunity for personnel to experiment and make mistakes in an atmosphere of security, promoting cross-curricular into departmental working groups, involving teachers in the design, providing the opportunities for networks for those teachers to think and reflect, devising and operating effective communications systems to ensure that teachers are kept informed institutionally, locally and nationally, providing LEA advisory support, demonstrating the total commitment of the LEA at senior officer and elected representative level, providing INSET at the point of identified need.' (17)

This quotation from Wigan emphasises the inter-relationship between school and LEA management; the latter is taken up in Section 8.

Most of the reports set out the time implications for doing the record of achievement in detail. For just the preparation of the summary document, an example is as follows:

'30–40 minutes to set an agenda for the summary statement

Tutor writing draft of summary, 20–40 minutes

Discussion of draft, 10–15 minutes
Total = 1¼ hours per student per tutor = 60 periods for a group of 30.

Secretarial time for a group of approximately 200 students will then need 70–100 hours of clerical support over a 20–week period' (1).

To this fairly typical description of the amount of time needed for the creation of summary statements, including the clerical processes involved, must be added the time necessary for in-service training — at the minimum of one day per teacher per year.

Others estimate 1½ days per teacher per year, which is not regarded as sufficient in some quarters. Time for the school co-ordinator generally persists at approximately one day per week, once the project is established. Time for accreditation, much of which would be 'hidden time' of LEA officers engaged in accreditation activities (11,13), is also needed. The 'one-off' costs are usually seen as the least problematic aspect of resourcing for the scheme and certainly the question of teacher time stands out as the critical issue. Although Dorset drew attention to the fact that improved teacher/pupil ratios did not always lead to a better records of achievement process, there is a feeling in more than one of the projects that more favourable staffing will be a necessary pre-condition for successful introduction of RoA.

'More time, more favourable staffing, without this commitment we will be playing at the scheme' (10).

'Without substantial government support it is unlikely that LEAs can provide schools with the support to make this a genuine reappraisal of the genuine curriculum content and delivery outcome. This is likely to make RoA marginal rather than central' (17). (Wigan identify elements of resource costs to include: provision of time and space for teachers, INSET, LEA advisory personnnel, increased ancillary support, materials and resources.)

The problem of sufficient time is further compounded by the range of initiatives currently affecting schools (8,10), as illustrated here:

'Apart from Records of Achievement other school activities such as TVEI, CPVE, GCSE are emphasising the need for progress reviews, counselling or appraisal. If such provision is to be a reality then either the very substantial resource implications have to be faced (e.g. in terms of releasing staff or reducing staff/pupil ratios) or there is a need drastically to redefine the working day in schools to enable these activities to take place' (10).

Essex strongly expressed the view that the standard of presentation and preparation that has currently been characteristic of their scheme cannot be maintained without increased resources.

It is clear from the above that the resources and time issues comprise two elements. The first is the practical element of actually providing the time and the expertise as well as the high quality materials that will make RoA practicable. The other is more a reflection of the degree of commitment that is seen to be put into the initiative by the school, the LEA and the Government. In this sense, increased resources are as much a statement of intent as a real practical aid to execution.

Institutions and schemes have varied in the way that they have sought to meet the time problem. Most have put important emphasis on the creation of more tutorial time. For example, in Essex the majority of institutions have:

'identified the need to increase the amount of time that was originally given to the exercise. Particularly to allow time for dialogue between student and teacher and therefore to encourage reflection and insight into the experiences being recorded. Several schools have therefore increased the allocation of time allowed by the 15–20 minutes registration period to include at least one full period of tutorial time per week' (5).

Lancashire suggested that a minimum of 2 × 35-minute periods per week for all staff is required to provide all students with a one-to-one review on a termly basis.

Essex made the distinction between younger pupils in years 1 to 3 and older ones in years 4 to 5, suggesting that only the latter would need a more formal review in each subject, requiring a minimum of 5 minutes per student per subject. Some teachers arranged such interviews in sessions in which pupils were engaged on individual tasks, while others requested time away from the timetable lessons for such discussion.

Many of the reports made specific mention of the new contract and its implications for RoA work. OCEA/Somerset for example suggested that the work has been managed quite differently under the new contract:

'In those schools which have received above-scale allocation of teachers and non-teacher time, work has been done within school hours, but in the others, which relied on voluntary effort, the scale of work has been greatly curtailed, due to the introduction of directed time. Future development will depend upon the priority given to RoA work and the demands of other priorities' (15).

Likewise, OCEA/Oxfordshire reported that the new contract had required many regular school meetings to be re-arranged during school time — involving some time being programmed for RoA. There is little doubt that the new contract has affected the number of meetings possible and attendance at those held outside the 1265 hours. Other schemes reported a continuing willingness to work on RoA outside the contract hours. ILEA has advised schools to allocate between 63 and 75 hours of directed time to assessment and LRA. This provides for meetings, conducting reviews, INSET, etc. During 1987/88, ILEA was able to allocate one teacher above authorised numbers to nearly every secondary school who was able to cover the lessons of staff working on some aspect of the development. This proved extremely successful, but clearly is not a level of funding that can be maintained. Overall, though, directed time has highlighted the role of school management in planning and developing INSET. The new contract ensures whole staff attendance at meetings and training days and allows the formation of smaller task groups with specific briefs. This is leading to thoughtful planning and efficient use of resources (13).

Whether this is true for schemes in general is not clear, but there is a feeling coming from a number of the reports that time problems are most successfully resolved when a redefinition of priorities in relation to the school's work as a whole is involved. Wigan provided perhaps the most explicit example of this in its commitment to a curriculum entitlement of which RoA are only a part. This point was also made in the negative by Essex who suggest that worries about time are in many cases arising from teachers not having adjusted their teaching approaches to accommodate novel assessment processes and not understanding how RoA may be integrated into their teaching practice.

An important distinction needs to be made between the formative and the summative process in this respect, however. The production of summary documents seems to have produced particularly acute time problems. Apart from student time, which in the case of ILEA was typically 25 hours spent on drafting statements, school office staff took about 20 minutes to type each statement. In Essex the time taken to produce a summary document was calculated as follows:

Subject-based summaries – 3 hours (10 at 18 minutes each)

General statement — 1 hour 10 minutes

Record of experience — 15 minutes

Personal statement — 10 minutes

Checking of document — 10 minutes

This represents a total time of 4 hours 45 minutes per pupil. For a year group of 240 pupils, this will represent a total time of 1140 hours.

Several schemes reported that when final documentation was being produced normal clerical assistance could not cope with demand and additional help had to be brought in, although the use of word processors had made up-dating easier (10,13,17). Ways in which schools in Essex streamlined and altered procedures were as follows:

'For example, some institutions are replanning their annual calendar . . . others are tackling the problem of the collation and checking of information needed for the record of achievement and ensuring that factual information especially, is held on computer and printed directly from there, whilst the difficulty relating to the time needed for typing the student's personal statement is being addressed, in some instances, by the students undertaking this task for themselves. Tutorial periods are also being extended to create more time for the preparation of the personal statements by students' (5).

Some schemes made little or no reference to the use of computers in their report, implying that they were not significant in the development work (e.g. 1, 16).

Broadly speaking, the functions that a computer can perform in relation to RoA include recording pupil and teacher objectives and so helping with subsequent reviews; collating assessments across the curriculum; and producing the summary document, both interim and final.

In general terms, the advantages of using computers are principally the hope that they will save time in easing the burden of teachers' work. The computer can also present a good quality summary document, though there is some worry about users wanting to see pupils' handwriting. The disadvantages are: the lack of expertise across the whole staff, which may result in the computing aspect being limited to a small number of teachers, militating against the whole-school nature of the scheme; the cost of hardware and software; the difficulty of providing for the range of different uses the computer may have in relation to a RoA (e.g. pupils inputting to their fomative record versus clerical staff producing the summative record); the problems of obtaining or developing suitable software; and, above all, the tendency towards the stereotyped comment bank approaches that appear to be an almost inevitable concomitant of using computers to save time.

There is some discussion in the reports about whether or not the computer can help with the formative process. This depended, in part, on whether machines were kept in the classroom or in some central location, and how data were fed in. There is reasonable hope that when the ILEA computer programme has been trialled it will contribute to a radical reduction in the amount of time needed to produce formative assessments (8). At some schools pupils were able to type their own drafts of summary records — a facility to which students responded well.

It is clear that the central issue in the effective use of computer is the time and expertise needed to develop and specify the rationale and the system of assessment rather than merely to construct a computer programme (3, 8, 17). The programme devised by ILEA (a scheme that employed a full-time computer programmer for the development phase) had to be flexible enough to make use of the various approaches of different schools, to make it easy for teachers to write their own course descriptions and assessment statements and to write additional free text. Furthermore it had to be easy to use for both teachers and students, and to be able to relate to the wide range of databases in use in schools and, indeed, the variety of printers in use, a point echoed in the WJEC report (19). It would appear that RoA developments may have been moving too quickly to allow ready computerisation e.g. the criteria and comment banks continually change. Some elements are relatively static while others require continual change.

Hardware seems to reflect a move from BBC machines to IBM clones or some other large memory machine. Software seems to be of a great variety. WJEC have many schools which have developed and marketed their own software. EMRAP trialled their system using commercial software and ILEA produced its own flexible programme which is currently being piloted. Many schemes experimented with the use of optical-mark readers which are particularly beneficial in cutting down time involved where comment banks of some kind are used (3,8,19). In the current stage of development, it might be possible that commercial software could be developed to provide more generally for RoA work; this might be a worthwhile area for central development. Meanwhile the provision of suitable equipment remains a problem. One way round the question of cost is that adopted by ILEA which had a central store of equipment such as optical-mark readers and laser printers for loan to schools when required.

Confidentiality did not appear to be a major issue in any of the schemes. Any problems in this respect seem to be amply compensated for by the learning benefits and other advantages of letting pupils become involved in constructing their own records.

Two other successful ingredients in the successful use of computers would appear to be INSET for teachers, and the provision of clerical assistance for the input of data and/or the typing of summary documents where provision is not at a level which allows pupils the major responsibility in this respect. In one scheme, at least, it is reported that teachers are too worried about the development as a whole even to begin to face the IT issues (3), and IT is clearly an area where training for both teaching and secretarial staff is needed. IT can flourish only when RoA are integrated into the life of the school, and are an extension of other computerised systems in its management and administration.

Overall, it is clear that the implications of using computers are both practical and educational. The disadvantages of mechanistic records need to be set against the potential for pupil control and learning on the educational side. On the administrative side, the disadvantage of limited and clumsy systems can be set against the considerable time-saving in both the storage and reproduction of records that computers can provide.

It is clear from the reports that the LEA needs to have a policy on INSET for RoA and to recognise that it is essential for their success. All schemes have already demonstrated needs for the resources for staff development, even if not all of these have been quantified. What has not been explicitly faced in some schemes is how the strategy will change when dissemination of RoA to all schools takes place (in some cases this is part of a general lack of thought about strategy for the next stage). Because Suffolk has already extended RoA to all schools, it has begun to identify new staff development needs but said nothing about how it would meet these. Nearly all the reports mentioned the resource implications of providing INSET across the whole LEA and recognised that provision would be likely to be substantially less than has been characteristic during the pilot phase. ILEA, for example, provided a minimum of one day of INSET support for 134 of the Authority's 141 mainstream secondary schools during 1987/8. In the future the activity of the development officers will continue in individual schools but probably not at the level of one day per week per school, which is now the norm. Another problem is that, under the LEA Training Grants Scheme (LEATGS), reaction at the moment of expressed need will be impossible; INSET provision will need to be identified at least one year in advance with detailed costings. Thus, in contrast to the ESG model of funding, there will be a tension with the notion of staff development as needs arise, a formula which has proved so successful in records of achievement development up to now. WJEC also felt that LEATGS could not meet the needs of RoA effectively, though links with GCSE training were recognised: in some schemes, explicitly-linked provision has been made.

Detection of INSET needs can be divided into:
 (i) reporting of exercises designed to find out participants' perceptions
 (ii) those identified by the project co-ordinators.

Under (i) Suffolk and WJEC, for example, administered Needs Assessment Questionnaires. Among the 163 Welsh respondents, interview techniques and assessment and recording techniques were the topics most in demand. Thirty-one per cent of the sample had not received INSET at all. Other needs frequently identified in the pilot schemes were: teacher–pupil reviews; pastoral and tutor skills; training of the school co-ordinator; and the provision of guidelines and manuals. Dorset also mentioned the importance of formative profiling becoming an integral part of initial teacher training as well as a focus for INSET. Several reports (e.g. 1, 3, 17) emphasised the need to help teachers move from a mechanistic, subjective and unsubstantiated approach to record-keeping in favour of the internalisation of a set of "educational, professional and ethical principles associated with achievement". (17)

INSET took a great variety of forms in the different schemes. Many (e.g. 1, 8, 16, 17) emphasised the whole-school approach. The most extreme example of this was in Wigan where the whole staff of one school, plus parents, governors and secretarial staff all went away for a residential weekend. In other schemes, such as ILEA where the emphasis has been for the development officer to support work ongoing within the school, INSET was often much more fragmented, working with small groups of teachers on particular issues. Another very important thrust has been the opportunity for co-ordinators to meet together from various schools to address common issues. Dorset, for example, had 'cluster groupings' to encourage school co-ordinators to share their experience and other teachers were drawn together across clusters for particular issues. Wigan saw the sharing of experience, not least through visits, as the most valuable INSET experience.

One model that seemed to have worked well was that devised in ILEA, which designated one of the five training days for each school to be used for the London Record of Achievement. An INSET pack was prepared for schools by the Central Assessment Team. The pack contained a possible structure for an

INSET day, information notes on the history of the LRA, extracts about assessment drawn from a variety of sources, issues for discussion groups, and prompts to support staff discussion on the future development in their school. This INSET programme reached every teacher and proved very effective in promoting organic development and INSET efforts. In future the days used will need to be staggered to facilitate the involvement of the Central Team in a wider number of schools. Furthermore, when a senior manager from the school worked with the Central Assessment Team in organising INSET this helped to enhance teachers' perception of the importance that the head and senior management team of a school were giving to the initiative and their consequent interest and motivation.

Other successful kinds of INSET mentioned included: residential meetings; conferences for heads; meetings of co-ordinators on special issues; and the provision of teacher fellows. Several schemes mentioned the growing involvement of the advisory and careers service which had, in some cases, been little involved in the past. In Dorset the LEA subject advisers were now using members of the County Assessment Unit to add an assessment and recording dimension in their INSET work. In other schemes such as ILEA and Wigan the advisers and inspectors had begun to take major responsibility for the development and co-ordination of the pilot work.

Some further general conclusions about key features on INSET can be identified from the reports. First, following the recognition of the importance of the school co-ordinator (see Section 4), INSET for co-ordinators and an organised programme enabling them to meet regularly and identify needs across institutions were important for maintaining both expertise and morale. Dorset gave this a particularly high priority, reflected in the plans to provide INSET for a school co-ordinator for every middle and primary school in 1988/89.

Secondly, school-based and school-focused INSET needs to be run by people who have themselves gone through the development process. In response to a school which has already been through the process of identifying its needs and is concerned to develop its own solutions, such INSET is best conducted under the management of an experienced co-ordinator, with teachers actively participating and given the time, support and the guidelines necessary to work through and address the particular issues of their own situation. The shortage of outside experts, and the difficulty such outsiders have in addressing the very specific needs of individual schools, suggest that this kind of approach is much less helpful. However, there is clearly a balance to be drawn between helping schools to develop on an individual school-focused basis and feeding in knowledge that is accruing to other schemes up and down the country and that may be of benefit to members of the development team in the school. This underlines the importance of the provision of workshop materials which can inform and support INSET activities.

Finally, the level of INSET provision is of critical importance. In several of the schemes INSET represented a considerable expenditure. Several project directors clearly felt that there is a direct relationship between the amount of provision made for INSET and the success of the scheme in establishing a truly organic and effective records of achievement initiative.

Suffolk coined the phrase 'the tightrope factor' to describe the balance between school flexibility and an LEA framework. The remark of one interviewee in the WJEC survey ('You have to allow some flexibility because you can never get two schools to do exactly the same thing') seemed to sum up the same thing. This model of 'bottom-up' development involving shared ideas (rather than an imposed common approach) reflects the belief that classroom teachers must be involved with the development. Few reports made an explicit link with the notion of accreditation principles as the framework for communality within this diversity although WJEC saw the accreditation principles as a strategy for holding schools together.

Generally, there appears to be a need both for leadership by the LEA and for support from it (8, 17), that is, the promotion of common and consistent policies and the provision of approriate support for teachers, meeting their perceived needs both in the short and long term. Thus, in Wigan, core team support covered financial, administrative, INSET, consultancy and personnel issues. The core team support model will require expansion to include the whole of the advisory service and LEA as RoA spreads to all schools.

The role of the inspector or adviser emerged as crucial in relation to both the leadership and the support functions. In OCEA/Oxfordshire, for example, a new advisory structure was introduced in January 1988, which assigned to each secondary school a general adviser. The OCEA accreditation system requires each school to draw up a plan involving self-review, target-setting and evaluation in consultation with its adviser. This gives the advisory service greater ownership of the RoA work in the LEA as well as providing in many cases a convenient focus for their generalist role.

The value of such high quality and aware advisory support to schools as a resource in their attempts at change was further underlined by OCEA/Coventry and was implicit in many other reports. ILEA inspectors have also taken on an increased role in relation to quality assurance through their normal programme of visits and evaluating schools' quinquennial reviews. An Inspectorate Steering Committee was established for the LRA and the Inspectorate and their advisory teams produced a number of booklets to help with profiling in, for example, English and computing. Members of the Central Assessment Team, inspectors and advisers are now working closely together.

The strategy of a central assessment team was adopted in several pilot schemes (e.g. 1, 8, 17). In Dorset, the team was directed by an Assessment and Examinations Policy Executive Group, in addition to the product advisory support group; there will also be a sub-group of the county curriculum working party to formulate the policy on assessment and recording in the light of proposals for the National Curriculum. In September 1987 an expanded Central Assessment Team was established to support implementation of the ILEA's policy on assessment under the umbrella of the London Record of Achievement. In such a development, which in some cases is the largest and most far-reaching ever undertaken by an LEA in terms of its management and curricular organisation implications, the collaboration between advisers and the core assessment team, as well as senior officers of the LEA and elected members, seems to be a vital ingredient in success. In addition, many LEAs have set up an RoA resources centre which provides another kind of resource for development work (8, 11, 17).

The development of RoA involves a number of different elements. For example, Dorset distinguished between structural planning (including the design of the scheme, development policies, plans for in-service etc.), the provision of structures for validation and accreditation, and the imposition of a detailed timetable — for example, for the issuing of summary records of achievement.

In summary, effective support by the LEA needs to include:

1. The provision of a minimal number of general guidelines or accreditation principles around which practise must be organised.

2. The provision of some kind of facility to provide both for quality assurance and for the continuing support of the development, most commonly through the inspectorate or advisory service.

3. The provision of resources, partly in relation to (2) above, but also more specific resources, such as computer hardware, materials, examples from other schools etc.

4. A responsibility for bringing schools together to share expertise.

8.2
Co-ordination

Perhaps the most important element of the LEA's role in relation to RoA is its responsibility to co-ordinate the range of pressures and initiatives currently impacting on schools. Not only does LEA co-ordination offer some solution to the overload of initiatives, but also helps consumers to have some kind of coherent product to respond to and thus assures the ultimate credibility of RoA themselves (e.g. 1, 15, 17).

However, a proposal for co-ordination raises a number of questions such as: Who manages the coherence within and between schools? How are the conflicting aspects of the different initiatives to be reconciled? Is it possible to avoid prioritisation among the initiatives? There is some indication that, although RoA can provide an umbrella for other initiatives, they may not always be seen as a top priority as compared to, say, GCSE or even TVEI. Some of the solutions suggested include:

- defining the common aims among the initiatives, within a common curriculum policy
- avoiding labelling development work as belonging to a particular initiative
- having one pilot school deal with several initiatives
- having a co-ordinating committee such as an inter-project group
- meetings of project co-ordinators
- INSET days for co-ordinators
- putting representatives of schools piloting other initiatives on the RoA Steering Committee at LEA level
- having a single RoA folder for all initiatives
- making sure that one officer is charged with checking the compatibility of decisions with other LEA commitments
- offering a classroom support unit which works in all schools to support any initiative
- having a shared closure day for INSET

Thus OCEA/Oxfordshire has appointed a co-ordinating adviser with responsibility for pupil assessment, school self-evaluation and teacher appraisal. Dorset also, from the start, integrated RoA development with curriculum development at county level, appointing five curriculum development officers in each of five key subject areas and co-ordinating GCSE and RoA by the secondment of seven teachers to provide INSET for GCSE alongside the RoA team. Dorset has now established the Dorset County Assessment Unit run by a director, whose appointment was seen as critical in bringing together the various initiatives such as CPVE, TVEI, GCSE, NVQ and, in the future, A and AS Level and national assessments.

However, the post-16 arena appears to present a particular problem in this respect, with RoA being constrained by the dictates of assessment procedures built into existing accredited courses such as those of BTEC and RSA, which may not reflect the more student-centred approaches pre-16 (17). Other reports noted the tension for staff on pre-vocational courses involved in both a college profile and the system of assessment and recording demanded by examination bodies such as CGLI, BTEC, and RSA: 'The problem is that we've got two systems in the college without the time to run even one.' (10)

While it is clearly possible and desirable for LEAs to co-ordinate the support and development associated with a number of cognate initiatives, there are nevertheless barriers to their complete integration. It is already clear that there is likely to be some profound tension in attempts to link TVEI and GCSE. GCSE, in particular, has been both a help and a hindrance to RoA. An example of the positive benefits are shown in this quotation:

'The onset of GCSE with its INSET has complemented the work of records of achievement and provided an added incentive for change in a number of departments. Even where departments have been slow to accept new ideas they have been engaged in presenting GCSE criteria in a manner which is meaningful to the pupils.' (5)

In contrast, RoA development work was a co-operative venture while GCSE was regarded as compulsory and unavoidable within the context of the teachers' dispute (15).

Thus at one level GCSE provided an impetus and a framework for many RoA schemes to develop techniques and procedures for delivering RoA within subjects. At another level it contributed to the erosion of goodwill and represented a further demand on time available which may have reduced the willingness of schools to undertake RoA in the fullest sense.

TVEI is referred to in many of the reports as an important parallel development with RoA in which cross-fertilisation of INSET and practice is very important:

'The elements of learning objectives, active learning strategies, appropriate approach to assessment, transferable or cross-curricular skills, progression and the involvement of the students themselves are common to both RoA and TVEI. The preparation and experiences of the staff involved are mutually supportive.' (10)

'Much of the work which has been done within Records of Achievement has provided an ideal platform for launching into the TVEI extension.' (5)

Mention was also made of the links between RoA and YTS assessment, suggesting that links between schools, colleges and approved training organisations can strengthen and develop this relationship further in the future. In many LEAs, the range of initiatives of this kind has given a strong impetus to the development of a County Record of Achievement and a County Curriculum and Assessment Policy. As already mentioned, this co-ordination was markedly facilitated when there was a central unit of some kind charged with networking developments across an LEA and indeed across a consortium of LEAs. The creation of a group involving representatives from RoA, Low Attaining Pupils Project, TVEI, CPVE and other similar initiatives, such as project co-ordinators, seconded teachers, advisers and other interested parties, has been a fairly general practice to provide for such co-ordination.

8.3
Context issues

8.3.1
GCSE

It is appropriate to move from a discussion of co-ordination as an explicit issue to the more general aspects of the contexts affecting RoA development. Clearly GCSE was one of the most important influences in this respect and has already been mentioned. Most of the comments suggested that GCSE had a beneficial effect, for example:

1. By encouraging and fostering new ways of teaching and assessing (e.g. 3, 8, 19).

2. By encouraging the direct use of GCSE criteria in subject assessments (e.g. 8, 19). Thus, for example, ILEA report that where grades were used on the RoA, school-based grades were beginning to be replaced by a grade that relates to GCSE criteria for students across the 11-16 age range. This led them to suggest that GCSE assessment criteria would influence the way that courses are taught from the first year.

3. By making time available (3). However, in at least one scheme, GCSE was felt to have pushed RoA into second place and to have swamped any effect that RoA might have been seen to have had (19). Part of the reason for this was the unfamiliarity with GCSE on the part of teachers, the time it took for them to do GCSE assessments and the demands of the syllabus which leaves teachers with too little time to engage in reviewing.

The relationship between GCSE and RoA was thus such that useful support came in both directions.

8.3.2

National Curriculum and assessment

The proposals for the National Curriculum and its associated assessment were referred to by a number of reports as creating a climate of uncertainty over the long-term future of RoA. For example:

'Until the long-term future of Records of Achievement is assured and until heads are clear about the likely requirements of other initiatives such as the National Curriculum, as regards use of time, schools are wary of sweeping changes in organization. So far little change has occurred in the way in which schools manage the timetable. Minor adjustment has occurred to allow for more tutoring time but no school has made drastic alterations.' (11)

8.3.3

Teachers' union action

Teachers' action was a constant theme in the accounts of the early development of most of the projects. In many cases it effectively turned a three-year project into a two-year one and meant that fewer summary documents than expected were issued, but its effect on morale was also important (e.g. 19). However, the legacy of teachers' action is no longer apparent.

Although relatively little was said in the reports about context issues, common themes are clear. A full advertising campaign to raise the general level of awareness and commitment is viewed as vital in educating both potential users and LEAs that the government is serious and whole-hearted about its commitment. It is too early to assess what the substantive impact of a national curriculum and assessment will be on RoA, but this again is an area that will need to be addressed with great care if the lessons of the pilot work as represented in these project reports are not to be lost.

8.4

Consortia

The role of consortia in relation to support and co-ordination provides an extra dimension, but the reports offered little new guidance about how to provide for co-ordination between initiatives. It would seem to be the case that integration across initiatives is essentially an LEA responsibility and that consortia, while providing the advantages of status and credibility in the eyes of users, a wider range of experience for teachers to draw upon and the other advantages, which are by now well rehearsed, may even exercise a constraint on the organic school and LEA evolution of new approaches to certification, because they impose a cross-LEA template of principles that some LEAs may have difficulty accommodating, given their policy on other related issues.

Most of the reports confined their discussion in this respect to parents on the one hand and employers on the other. A few deal with users more generally, for example Dorset reported that most school governors appear to be in favour of RoA. In Suffolk there was a mixed reaction from FE and sixth forms, with one institution using RoA and two others, not. Heads and co-ordinators in the WJEC scheme thought the response of YTS organizers was positive. Also in the WJEC, interviews with careers service representatives revealed that they used summary documents as a source for reference and the formative process as a preparation for job interviews. However, they were worried about the timing of the summary documents. In ILEA, college staff stressed the importance of students receiving advice and practice at school to help them make the best use of their London Record of Achievement at interview, and to select samples of work that were appropriate for different interviews. This point was also made by Wigan, who suggested that students need more help on how to use the RoA in applying for a job, selecting the appropriate sections for interview and preparing a CV from the information. They also made the important distinction between informing and involving users.

9.1
Employers

Employers were involved in validation boards in Dorset, and in OCEA/Coventry employers are involved in RoA work via Accreditation Support Groups, school associations and a school validation board. This supports the principles involved in partnership and it has been suggested that industry would welcome more involvement at the formative stage through, for example, work experience schemes. In ILEA, the London Education Business Partnership was established to foster links between London employers and the ILEA. A group of personnel managers from major London firms, chaired by the personnel director of United Biscuits, was established to prepare guidelines for teachers, students and employers on what information in the LRA might be of value to the employers. The group included representatives from BP, IBM, Marks & Spencer and Whitbread.

Generally, the response by employers to RoA has been positive although several projects (e.g. 17,19) mentioned the need for something to be done nationally, for example through the CBI, to heighten awareness of RoA. Suffolk too emphasised that publicity will be needed for a number of years until RoA are established on a national scale as well as on a local scale through, for example, the existing school/industry links. Local publicity in Dorset included the appointment of the Information Officer to inform employers through, for example, the creation of an exhibition for use with employers and schools. ILEA is producing two videos, one for an audience of teachers, the other for an audience of students, parents and employers for this purpose. Only 47 out of a potential 1,000 turned up to an Employers' Conference in Wales and generally the involvement in development appears to be poor, perhaps because some LEAs think it is inappropriate. However, against this, 19 per cent of employers responding to a questionnaire in Wales reported that they had been consulted, but 60 per cent said they would have liked to have been consulted. Wigan suggested a number of specific strategies for a national publicity campaign including: the use of employer network services, business associations, clubs, chambers of commerce and trade, YTS/JTS network meetings, meetings of professional bodies such as the IPM, Rotary Clubs, accredited training centres, and enterprise schemes, as well as mailshots using available publicity, tapping into exhibitions, educating job centre staff and careers service personnel, the use of local radio and so on. Finally, they suggested contacting those responsible for recruitment at the top of organisations and large companies. In relation to this, several schemes mentioned the need to standardise the currency value through some form of accreditation, so that the quality of the RoA from different schools could be regulated (5,8,17). Wigan employers also mentioned worries about the lack of standardisation even within a school as well as across institutions.

It is very difficult to generalise about employer reaction to RoA since this is, in most cases, qualified according to what the RoA contains and how this response was studied. The WJEC survey suggested that 50 per cent of employers rated 'personal qualities' and 48 per cent 'subject skills' as the most useful sections. Seventy-three per cent thought it could replace a confidential reference. This was not the case in Suffolk, where most employers felt there was too little information, but this perhaps reflects the more pastoral nature of the Suffolk RoA at present. In Wigan there has been a special effort to include employers from the public sector as well as the private sector, on the assumption that their requirements might be different. The only common requirement from the employers studied was the desire for externally accredited information about maths and English; as in other schemes, 'personal qualities' were high on the agenda too, for example, 'inter-personal skills', 'problem-solving skills', 'adaptability', 'punctuality and attendance', 'breadth of experience and achievement', from both within and outside the curriculum. In particular they mentioned 'self assessment by pupils' as being a valuable preparation for job appraisal schemes in the world of work. In Wigan, the ideal design was investigated in some depth by inviting delegates to work in groups of two or three on the records of individual pupils that they would interview the following day on a two-day INSET course. The exercise was designed to discover how much and what type of information was manageable, acceptable and useful at the interview stage. Problems that arose included the use of over-zealous adjectives, shortage of simple facts relevant to the career, and the need for a summary sheet as well as marking and grading systems requiring simplification and explanation. Some employers wanted to know more about 'bad' points too, and some evidence of how weaknesses have been overcome. There was no agreement about the use of estimated or projected grades but they did feel a summary sheet for each year in the folder would be helpful to help cut down the amount of paper. In Essex too, many employers felt they would like to see some evidence of individual weaknesses although many agreed that the summary record might not be the place for this; there, though, there is a continuing commitment to providing confidential references which makes this point much less problematic. In Essex, the remark made by a major employer to one pupil was regarded as typical:

'This is really tremendous, it will at least double your chances of getting a job, it's just the sort of thing employers are looking for. If only all the young people who came here could have one of these'.

In an evaluation study carried out in Essex, copies of three documents were shown to some of the employers (7). The pupils represented in the records covered a cross-section of ability and achievement. The employers had little difficulty in placing students in the rank order perceived by the school – this may be both a strength and a weakness. Other comments from Essex employers included:

'I think it's great but there's still a need for a confidential reference.'

'These should ensure that we give prospective employees a fair chance to show their strengths at interview.'

'I like the idea of a standard format, I shall soon get to know what I'm looking for.'

'These documents could give your students a most unfair advantage!'

Wigan provided an example of a student returning from an interview at Ford's Halewood plant. He was specifically asked to report back to his headteacher that personnel staff had been most impressed with the document; they had not previously encountered anything which gave them the kind of detail they found and they wanted to send a message of thanks and encouragement back to the school. In ILEA, employers stressed their interest in extra-curricular activities and work experience, the students' plans and ambitions and key achievements over the last five years. They attached considerable significance to the choice of samples of work made by the students for the portfolio.

Naturally all schemes did much to inform parents using meetings (e.g. 2,3,16,19), letters and publicity sheets (e.g. 3,17) and surveys (e.g. 3,19). The results of the WJEC survey are not reassuring, however:

Knowledge about your child's RoA — 74 per cent YES; 26 per cent NO

Knowledge about the pilot scheme — 60 per cent YES; 40 per cent NO

Enough information — 38 per cent YES; 62 per cent NO

Attended meetings — 21 per cent YES; 79 per cent NO

Speak to form tutor — 23 per cent YES; 77 per cent NO

Fewer claims were made about parental involvement in RoA although, in the case of the parent's own child, this can obviously come through receiving and commenting on an interim or summative report. The WJEC survey showed a low involvement and a surprising reluctance even to be involved with the child's own RoA:

Involvement in RoA — 14 per cent YES; 86 per cent NO

Should be involved with child's RoA — 45 per cent YES; 55 per cent NO

However, it must be remembered that parental definitions of involvement may differ from parent to parent and school to school. What they do receive they seem to be happy with; for example 80 per cent were 'satisfied' and 84 per cent 'think it's fair' (19). In EMRAP/Northamptonshire, some parents wanted negative comments on pupils as a spur to positive achievement, while others wanted GCSE grade predictions. In the WJEC survey 72 per cent of parents felt that the 'personal qualities' section was the most useful, 45 per cent 'exam results', 39 per cent 'personal interests and achievements' and 32 per cent 'form tutor comments'. In Essex, responses came from 76 out of 170 first-year parents consulted at one school. Thirty-seven were wholly in favour of the new-style report and many thanks were given, 13 were generally in favour but had minor criticisms or suggestions and one was wholly against – thinking the old-style report was better; a further 25 commented only on their own child's report. Reactions to another school's change in its reporting system included:

'It has proved very useful when linked with the parent partnership scheme, because it enables me to see the areas of weakness and provide extra work at home which is done with the parents.'

'Allowing parents to comment lets you know how the student copes with their work at home' (10).

OCEA/Leicestershire reported that parents have been pleased by the increased information in school reports and, when involved in some of the same formative processes as the students, have been enthusiastic. Those who worried that their children were guinea pigs were reassured and felt there was a lot of positive information coming from this approach.

In Essex there was some concern among parents about how standards attained could be judged from the statements in the comment bank used. Parents who might only have a set of positive statements on a report took the view that some other information was needed in order to judge progress and effort. Wigan raised the important caution of the need to discover how pupils feel about being monitored at home and at school, and about the conflict between school regulations, which require a school to produce information about a child's work on request by the parents, and the ethos of RoA in which the student decides who is to have access to information contained in the record.

There is some confusion in the way in which these terms were used in the reports, in that some reports used 'validation' where others would have used 'accreditation'. Part of this confusion is due to the fact that the processes themselves are linked. An example of the confusion is the statement issued by Nottinghamshire concerning the accreditation of summary documents:

'Nottinghamshire County Council Education Committee is satisfied that the pupil named in this record of achievement was involved in all the processes from which it was compiled and that any statements made about the pupil's achievements are based on real evidence' (3).

This is more like a statement arising from a validation exercise than from an accreditation process.

A similar confusion was evident in the Wigan report, as follows:

'A validating statement is either included within the summative record of achievement or on the folder containing the summative document. This will be either NPRA or Wigan LEA. Validation is through the ground rules of NPRA.'

'The validity of Record of Achievement can best be judged in terms of LEA policy and the set of principles which underpin the content and delivery of the curriculum which provides a reference point by which to judge the validity of both the content and the process inherent in the Record of Achievement that is produced.'

There is an inevitable overlap between the validation of the particular courses or programmes of study and the accreditation of institutions to issue records of achievement with a particular external endorsement. The difference between the two processes is perhaps best illustrated by an example of each. The EMRAP scheme for accreditation, for example, which was used in Derbyshire, Northamptonshire and Lincolnshire (Nottinghamshire had its own scheme) is basically a two-stage process. Stage I involves a local group (an officer from the LEA, a representative of the local consortium, and a fellow practitioner) working with the school in a dialogue designed to clarify the principles of EMRAP and the school's processes in preparation for Stage II. Stage II then consists of a presentation to a full accreditation group (i.e. the local group plus a representative from the regional consortium, the examination board and a representative from another county), and a discussion with them leading to an agreement on targets. This in turn leads to the final accreditation decision. Development follows, with INSET being provided as required and a review taking place sometime later. From the point of decision a school can issue a record of achievement folder with the EMRAP endorsement.

In Dorset, by contrast, the emphasis was on validation boards. These were established by the governing body of a school and consisted of members of the teaching staff, members of the governing body, parents of students in the institutions, representatives of the LEA, employers in the local community, representatives of other schools in the county and other members co-opted as appropriate. The role of the validation board was to assist the institution in the development work and to assist in the process of awarding certification to the individual records of achievement:

'The validation board is involved in issues of curriculum, methods of learning and assessments. The teachers in the institution demonstrate to the validation board that the common principles have been upheld and are being promoted, and that a regular process of review is being undertaken in terms of four stages:

1. Declaring intentions in relation to curriculum, teaching methods and learning styles, assessment and recording, administrative arrangements.

2. Granting approval, in which the board examines the intention of the institution in relation to county policy for curriculum assessment recording and Records of Achievement. The board then decides either to give approval or indicate what further needs to be done in order to gain such approval.

3. The ongoing monitoring position, and

4. the granting of approval to issue the award' (1).

The composition of the boards raises the issue of whether it is a professional or a consumer responsibility to grant approval for the educative process. The rationale of the procedure reflects the concern that the processes that a school provides for its students are critical in helping the youngster achieve and demonstrate what she 'knows, understands and can do'. In Dorset, validation boards have an important part to play in influencing the educational process. For example, a PE department in one Dorset school set up a series of problem-solving tasks in which pupils' leadership qualities were assessed and recorded in response to a validation board's view that 'the development of leadership qualities should be part of the curriculum.'

Nottinghamshire's procedure was a variant of EMRAP's and offered an interesting combination of these two approaches in its four-stage process. The first involves the formation of an accreditation panel comprising education officers, an adviser or inspector and a head. Stage II is a presentation by the school to the panel of its development work and practice in relation to eight accreditation principles. Stage III involves a validation group made up of advisers and inspectors, led by the panel adviser representative, who collect evidence to present to the panel; Stage IV is the decision by the panel. Stage III, which is essentially one of validation, involves the advisers and inspectors monitoring the work of four pupils in some detail.

The confusion between validation and accreditation is further reflected by the situation in Dorset which is now becoming part of the Southern Partnership for the Accreditation of Records of Achievement, the aim of which is to provide regional support to individual LEAs in terms of 10 accreditation principles. The hope is that SPARA will facilitate linkages between RoA and other current initiatives such as GCSE, TVEI, modules etc., provide for economies of scale in terms of the implementation of national RoA guidelines, help to give RoA national and international credibility, and provide support for INSET and evaluation work.

It seems clear from these examples that schemes feel the need for a process which both allows some kind of external lever on the curriculum which will ensure that the latter, in terms of its processes and content, are in line with the overall thrust of the LEA and, at the same time, conveys some kind of external approval designed to reinforce a number of basic RoA principles and to support the status and credibility of the document in the wider world.

Some further issues that arise from this include:

1. Is it the quality of the process that is judged or just its existence? (For example, a school may provide for a 30-minute one-to-one review for each pupil but the existence of such a review says nothing by itself about the quality or utility of it.)

2. The composition of the validation/accrediting individual group panel: should it consist only of professionals or a blend of professionals and consumers? Are local advisers sufficiently independent to do this job, and what is the trade-off between having a developmental role and having a judgemental role?

3. Cost effectiveness and use of time: the Southern Partnership for the Accreditation of Records of Achievement proposes a scheme which operates on the mutual support of LEAs but nevertheless there are clearly considerable time implications for such procedures as well as the costs where an examination board is involved.

4. What counts as meeting the accreditation criteria? At what point can a panel say, 'The principles have been satisfied'? The answers depend on the extent to which accreditation is seen as developmental or inspectorial (3). Furthermore, the criteria may be easier to use and interpret when they are more operational (19).

Other questions raised concerned how often an institutional review should be carried out and the difficulty of judging process in one-off visits. All these issues will have to be addressed by RANSC in considering the nature and the implementation of national guidelines. Further insights into this difficult area will only gradually become apparent as accreditation of school processes becomes more widespread. However, there is clearly considerable experience that currently exists, particularly in higher education, and it would be sensible in the light of the pilot schemes' experience to consult explicitly with some such bodies for advice in relation to some of these detailed practical questions. The role of examination boards in relation to validation and accreditation still seems to be problematic. It ranges from simply selling the right to use their imprimatur on summative documents (by devolving the responsibility for accreditation to the LEA or consortium) through to schemes in which the examination board is the co-ordinator of the development work, as in Wales, and schemes where an examination board has a negligible role except in the provision for validation or accreditation for individual parts of the record, such as units or graded assessment (Wigan and ILEA). Suffolk, by contrast, is committed to the idea of self-evaluation of the summary document as an adequate presentation of the person it concerns. This view is further supported by the ILEA report which suggests that employers want no further validation than their own experience of using summary documents.

A final aspect of validation concerns the authentication of individual pupil entries on their record, particularly regarding extra-curricular activities and experiences which many schemes require to be validated by an adult (e.g. 3,19). WJEC was also concerned about the cross-validation of teachers' assessments of personal qualities, and moderation of a standardised checklist produced by the consortium is advocated to ensure fairness of assessments across the consortium, but this suggestion has not been taken up elsewhere.

In sum, the key questions would appear to be:

1. Is the content of the record, formative and summative, as true (i.e. valid) as it should be? What procedures are necessary to ensure that this is the case in the construction of a record of achievement?

2. Is the school providing the right kind of curricular experience in terms of teaching and assessment to allow all pupils the opportunity to experience success on a wide range of fronts and the opportunity to record that success (is the process valid)?

3. What measures need to be introduced to assure potential consumers of the status, generalisability, integrity and comparability of records of achievement?

Anything that is said about the general impact of the attempt to develop a RoA scheme must be set in the context of the proviso that several reports made, namely that it is very difficult to separate out the independent effect of RoA from among a host of other initiatives. WJEC in particular point to the impact of GCSE which swamped the RoA effects. That said, most of the reports refer to some change in teaching and learning styles involving the clarification of aims and objectives and the benefits of sharing targets with pupils, giving them something to aim for. All the LEAs participating in OCEA referred to this change, for example:

'There is no doubt that OCEA has contributed to changing styles in teaching practice. These changes have in themselves, in some instances, brought about a fundamental change in teacher/student relationships. It would appear that these two factors are crucial to student motivation, but they are also evident as factors related to a number of initiatives such as TVEI, CPVE and in work throughout schools' (12).

'Records of Achievement has forced us to consider how our curriculum is delivered to the pupils, notably: the purposes, the content, the methods, the assessment of progress and the achievement of shared implicit goals. This has perhaps made lessons more varied, more pupil-centred and more relevant and hence pupils appear more responsive and motivation is better. Better teaching produces a better response in pupils, but is this due to OCEA or to the enthusiasm engendered by better resourcing and greatly increased level of INSET?' (15).

In Dorset, similar sentiments underpinned the assertion that both motivation and behaviour had improved as a result of the RoA scheme. There had also been a potentially fundamental shift in focus from a view of the curriculum as what the school provides to a view of the curriculum as what the student takes away.

But this has not come easily. In the words of a Somerset teacher,

'Trialling materials and new strategies has often unsettled our confidence as teachers. Some of us have found this aspect quite offputting. But the new perception of how much the process of learning and assessing can be shared has been worth every bit of the shake-up and has steadily grown in our work. Most of all, perhaps, our development in learning and teaching styles has been in exploring the ways we can each, in our own ways, in our own departments, manage this new range of assessment techniques. We are better at our job and are confident our students get on better with theirs. The sudden shafts of sunlight over the moors of struggle have made much of the uncertainty worthwhile' (15).

Very little evidence is, or can be, offered for these kinds of perceptions and all that can be concluded is that at least some of the teachers in some of the pilot schemes have experienced this kind of increase in confidence and morale, and perception of the beneficial effects of RoA. How far this is a general response among teachers is still difficult to judge from the reports. Only the WJEC report referred explicitly to the problem of hostile teachers and what to do about them. The existence of more than a few of them can severely limit the potential of the project. The WJEC report also provided an instructive contrast to the perspectives quoted above by offering data on the extent to which teachers perceived their teaching to have been affected. Forty-three per cent of the sample felt that their teaching style had been affected only a little and 46 per cent not at all. Forty-six per cent also felt that the **content** of their teaching had been affected not at all.

Other effects of the RoA scheme beyond actual teaching practice in the classroom include an improvement in student/teacher relationships (e.g. 3,5,16,19), and the beneficial effects on departmental discussion, with the focus of such meetings more on professional items and less on administrative business, leading to more collaborative work between departments and more team spirit (13).

Several schemes (e.g. 3,5,11,19) refer to improved recording and reporting to parents and improved relationships with them as a result of RoA work. Related to this is the ILEA report of a growth in certificates that recognise specific achievements and experiences and an extended use of classroom and school displays and samples and photographs of good work.

Summary It is possible to summarise briefly the general impact as perceived in terms of the pros and cons of the RoA.

Cons A continuing and obvious problem is that of time: time to devise the assessments, to manage them in the classroom, and to negotiate with students. Time is, of course, related to resources and the need for the status of RoA to be clearly settled, in that ambiguity in this respect tends to be reflected in poor resources, commitment and morale. Other problems mentioned explicitly include the effects of teachers' action, the uncertainty about the extent to which new initiatives may make conflicting as well as additional demands, and simply the volume of demands of such initiatives.

Other issues relate specifically to the nature of the ESG pilot project, such as concern about the wide range of interpretations that existed at the beginning of the project as to what records of achievement were about, concern about the provision of adequate resources beyond the lifespan of the project, and worry about the shortness of the initial project in which three years has not proved sufficient time to allow new practices to permeate every teaching group in schools.

Pros There is a widespread belief that the best practice in records of achievement has led to better relationships between teachers and pupils, and an extension of the partnership with parents. The amount and quality of dialogue have improved, and pupils have begun to take more responsibility for their work, leading to improved learning and more active involvement. The curriculum has also changed for the better in two respects: first, in the clarification and communication of learning objectives and, second, in the integration of the academic curriculum and personal and social education. Although not all pupils appear to benefit equally, there is evidence that some of the lower attainers gain from the experience. Much less can be said about the value and credibility of the summary record: while many remain convinced of its value, the evidence is not yet available.

PART THREE

Synthesis

by Patricia Broadfoot

Contents

According to the 1984 DES/WO policy statement, records of achievement have four purposes which might be paraphrased as follows:

(i) to recognise, acknowledge and give credit for what pupils have achieved and experienced in a variety of different ways.

(ii) to provide encouragement and increase pupils' awareness of their strengths, weaknesses and opportunities in order to enhance motivation and personal development.

(iii) to help schools support and encourage the development of pupils' diverse talents and skills through a consideration of how well their curriculum, teaching and organisation serve these ends.

(iv) to provide a summary document for school-leavers which will be valued by potential users for the wide picture of a young person's qualities and achievements that it portrays.

The Record of Achievement National Steering Committee's Interim Report of November 1987 was able to state that:

'while progress has been made and ideas have developed, the principle ('that boys and girls who stay at school until they are 16 may reasonably look for some record of achievement when they leave') remains unchallenged, as do the four purposes of records of achievement identified in the 1984 policy statement'.

The sustained and widespread support that continues to be expressed for the original policy document's articulation of the purposes of records of achievement suggests that whilst the extensive development work of the last three years may have refined and deepened our understanding of the issues associated with records of achievement, it has not in any way suggested that the original conception of the rationale for records of achievement was either inadequate or misconceived. There is rather a continuing concensus that records of achievement are an important and desirable innovation which can help significantly in meeting a range of needs that are becoming increasingly apparent both within the educational process and outside it. The fact that there has been no serious challenge to the policy itself reflects a situation in which both insiders and outsiders are convinced of its merits.

It is therefore appropriate that this same rationale should have formed the starting point for the national evaluation, the framework for data-collection and the focus for its final report. One of the first tasks of the PRAISE project was therefore to break down the original policy document into questions which could guide data-collection and analysis (see Document 3).

In our interim reports (1986, 1987) these questions were transformed and extended to provide the 'map' of issues associated with records of achievement about which we felt our data could provide some insights. In this final report it is our aim to provide the most comprehensive range of insights into records of achievement issues that three years of data-collection and documentary analysis can provide.

The fulfilment of this purpose however goes beyond the presentation of a summation of our findings. It must also include an attempt to **conceptualise** what the **realisation** of the four purposes of records of achievement would look like in practice. Or, to put it another way, how, and to what extent, it may be demonstrated that records of achievement 'work' in the way intended. What would constitute evidence that the various elements of records of achievement policy, most recently expressed in the RANSC Interim Report, are justified?

The preceding parts of this report therefore present a range of evidence and analyses designed to address such policy questions. As might be expected with such a complex innovation there are few areas in which our findings are unambiguous and do not require some element of qualification or reservation. Although we can speak with some confidence on certain issues, many others remain unresolved and we can offer little in the way of clear pointers to success. Still other questions remain essentially unexplored, largely because the pilot schemes typically had not reached the necessary stage in their development to provide evidence on which we could base any evaluation during our data-collection phase. In both these latter cases, further research is required and is indeed being supported by the DES/Welsh Office as an extension to the original evaluation project.

Our aim therefore in the report as a whole has been to deepen understanding of the issues, as much as to make generalisations about practice; to guide policy-making rather than give a blanket affirmation of success. To this end we have tried to include illustrative material which illuminates **the issue itself**, rather that what is necessarily typical or desirable.

The focus of this final part is rather different however. Whilst recognising that our insights must be, as yet, far from complete, our aim is to present in a more synoptic form the conclusions we have been able to reach about how far records of achievement schemes in practice can and do live up to the purposes which have informed their introduction. Thus this final part of the report is divided into two main sections. The first section addresses the question of whether, and in what ways, records of achievement 'work'. It evaluates how far the evidence that exists testifies to the success of records of achievement in meeting their avowed purposes. The second section addresses the necessary amplification of these very general findings in terms of conditions which, in practice, are likely to lead to such success.

1 How far do records of achievement appear to fulfil the purposes of the 1984 policy statement?

Attempting to provide an answer to this question quickly reveals that the four purposes of records of achievement identified at the beginning of this chapter are not of the same order. In particular, purposes One and Four are a great deal more specific that Two and Three and can therefore be dealt with relatively briefly.

In relation to Purpose One our evidence confirms that all the schools studied are fulfilling this purpose. As section 1.1 of Part One of this report sets out, all the pilot schools studied gave credit for pupils' achievements and experiences in a variety of different ways if not always across the whole curriculum. Having said this, deep and difficult questions clearly remain — as section 1.1 in Part One notes — about which achievements and experiences should be included and even how these terms should be defined. There still exists little consensus about the nature of cross-curricular skills, for example, or about the definition and assessment of personal qualities or skills. Schools are still not clear where the embrace of records of achievement should legitimately end and what the remit 'to address the totality of the pupil's achievements and experiences in and outside the classroom' (RANSC *Interim Report* paragraph 16) should mean in practice. Perhaps the most problematic issue however is **how** schools seek to do this, because our evidence suggests that it is the **way** in which such recognition is afforded, rather than its existence per se, that is crucial to the fulfilment of Purposes Two and Three as we argue below.

1.1 Purpose One: The recognition of achievement

The fulfilment of at least part of Purpose Four — the provision of a summary document for school-leavers — can, like Purpose One, be quite categorically confirmed by our data. Despite the fact that some schools regard the issuing of a summary document as of relatively minor importance compared to the formative process leading up to it, schools and schemes nevertheless appear to accept that they will provide a summary record of achievement for all school leavers. Even those schools which had yet to reach this stage during our field work had clear plans for doing so.

1.2 Purpose Four: The provision of a summary document

Necessarily more problematic is the question of the value such records of achievement have for potential users. Although some research has been undertaken to elicit employers' views in this respect[1], the relatively small number of such documents issued so far makes any such study inevitably artificial. It is impossible to predict how much value potential users will accord to records of achievement if and when they become a normal part of the educational landscape. At this stage all that can be said is that the signs are encouraging, and that there are many documented instances where users have responded both to the **idea** of records of achievement and to individual examples which school leavers have shown them.

Nevertheless, it is too early to say with any certainty whether the procedures currently being developed will meet users' expressed need for particular kinds of information which have clear credibility. For example in its response to the RANSC interim report the Further Education Unit — a body with a great deal of experience in this area — raised a number of questions about the extent to which records of achievement 'can **meaningfully** and **reliably** indicate a student's degree of attainment in specific areas' so that 'for students who are trying to progress to a further course of study or employment, RoA do not have the same currency as passes in exams or entrance tests.' Although there is no evidence that record of achievement assessments are any less reliable than examination grades there is some indication that they should indeed be more meaningful. Such worries nevertheless pose a variety of questions about the way in which evidence for the leaving document is collected and summarised; how it is

---

1. See for example, the NFER pilot study of employers' reactions and the 'Robert Arthur Essex' study conducted for Essex LEA by the Industrial Society.

presented; and the kind of arrangements that are made to give the final product credibility. The evidence we present in earlier parts of this report points mainly to a great variety in all three of these respects and no very clear way of evaluating their relative merits at the present time. It is clear that RANSC will need to consider carefully the degree of national standardisation of such documents that might be desirable to ensure a useful and credible report that will be of value to users and hence, just as importantly, to the pupils themselves. It is also clear that further evaluation of user response to records of achievement — such as that currently being conducted by NFER — will be an important element in choosing the optimum balance between autonomy and standardisation in this respect, and hence in the success of records of achievement as a whole. If the records are not valued by users, this will seriously detract from their perceived value for many pupils despite the sense of satisfaction that a significant number of pupils feel in simply having such a statement of achievement at the end of their schooling. The close link between the perceived utility of records of achievement in the market-place and pupil motivation which our data point to, suggests that once again, the **way** in which Purpose Four is met is also crucial to the fulfilment of Purposes Two and Three.

In order to inject a fresh perspective into the analysis and to provide some cross-validation of the original conceptualisation of records of achievement purposes, the discussion that follows is structured according to the rationale of the records of achievement process itself, starting with pupils, whose response is the raison d'etre of the initiative, and moving out in concentric circles to include the successively widening concerns of teachers and users.

1.3
Purpose Two: Enhancing pupils' motivation and development

Section 1.2.5 of Part One identifies two ways in which records of achievement may affect pupils' motivation.

'First, the processes of recording aim to make explicit many aspects of teaching and learning which have traditionally been left unstated, including purposes, objectives, short-term targets, and so on. The passive and dependent attitudes of pupils towards their progress and development at school have generally been blamed for poor motivation and under-achievement. Armed with clearer notions of what teaching and learning are about, and a language with which to think and talk about those notions, it has been hoped that pupils will be able to take over more of the responsibility for their learning. Second, the processes of recording aim to provide pupils with regular, detailed, positive and individualised feedback on their progress and development; motivation would thus be boosted by pupils' greater sense of success and accomplishment.'

Part Two of this report consistently reiterates how hard it is to **detect** changes in pupil attitude that can be directly linked to records of achievement and even harder to **detect** evidence of enhanced pupil development, progress and awareness of strengths, weaknesses and opportunities which have records of achievement as an explicit cause. Such bilateral relationships of cause and effect are unlikely in any educational context and certainly would require more long-term research to elicit changes in them. At a time when many other initiatives are impinging on schools such as the GCSE examination or Technical and Vocational Education Initiatives, it is often impossible to disentangle their separate impact. Furthermore such evidence points to the need to distinguish between changes in pupils' **attitudes** brought about by records of achievement and changes which records of achievement may have brought about in pupils' actual **competencies** — academic and personal.

1.3.1
Changes in pupils' attitudes

On the first count we can nevertheless point with some confidence to evidence that many pupils have found the opportunity of talking with their teachers on a one-to-one basis about achievements, experiences, needs and appropriate future targets a rewarding and helpful experience that has had a positive effect on their motivation.

How far do records of achievement appear to fulfil the purposes of the 1984 policy statement?

'We also have enough evidence to support the conclusion that pupils in our case study schools have grown in self-awareness and in their ability to reflect more acutely on their academic progress and personal development.'

'. . . records of achievement processes have provided considerable scope for the setting of short-term targets.'

Equally we have evidence from Part Two of our report that

'Pupils' motivation is significantly improved to the extent that they are given responsibility for their learning and are involved in reviewing and assessment. Furthermore, some schools report that pupils are becoming more confident and able to challenge, often in a very mature manner as a direct result of records of achievement. (Wigan)

Essex reported improved student self-esteem and higher expectations by both teachers and pupils and that students' learning strategies are becoming more autonomous and their capacity to engage in self-assessment and review is increasing as they master the assessment 'language'.

It is clear from this evidence that it is changes in classroom approaches and relations which underpin the impact on pupils. This is likely in turn to be a reflection of the extent to which teachers are genuinely attempting to institute not just a 'paper and pencil' exercise but a different approach to teaching and learning itself. It would appear that Purpose Two of the policy statement, like Purposes One and Four already discussed, also hinges on the fulfilment of Purpose Three.

It is also important to distinguish further between the different kinds of pupil motivation that appear to be associated with records of achievement. Some motivation, especially for girls, appears to be intrinsic and **related to the recording process** itself in, for example, 'planning for nostalgia' as referred to in 1.2.1(b) of Part One. This motivation needs to be distinguished from that which is not directed at the records of achievement process itself but rather towards learning more generally although it is in part at least a **product** of the records of achievement process. It is the changes in the teaching-learning relationship itself which are critical in this latter aspect of motivation — changes which may be far more pervasive than anything that can be directly attributable to records of achievement procedures. It would seem then that the impact of records of achievement must be further distinguished as that associated **directly** with the **procedures** involved — such as recording, reviewing and reporting — and that associated more **indirectly** with records of achievement **principles**, such as pupils taking responsibility for their own learning; the setting of clear objectives; mastery in learning and regular feedback.

(a) Intrinsic versus extrinsic motivation

By contrast, a potentially powerful source of extrinsic motivation is the record of achievement itself. Although relatively few of these documents had been issued in schools within the pilot schemes, by the close of data collection, there are nevertheless strong indications where they have been available that pupils are very positive in their response to them. In some cases pupils appear to regret that they did not give more effort to their preparation. It seems likely that to the extent that such records of achievement gain credibility and status in the market-place and are also intrinsically pleasing to their owners, they will provide a much stronger source of motivation in the future than we have been able to record up to now, as successive year-groups of pupils become more and more familiar with both process and product.

The realisation of this goal is still problematic, however. This reflects a certain innate tension between Purpose Two and Purpose Four of the policy statement. As is now widely recognised, schools and teachers may find themselves torn between what they know the market-place will value and what they themselves think is desirable in relation to records of achievement, where these are different. The problem of moving from 'formative' to 'summative' is for this reason well recognised, as is the difficulty of respecting pupils' rights to include whatever is important to them in their record even when it is clear some entries may not serve their interests. At the root of such dilemmas is the fundamental question of whether the motivational impact looked for, from records of achievement, is intrinsic or extrinsic. Intrinsic motivation includes the building of self-esteem; the setting of realistic targets; the sharing of responsibility and the recognition of the individual that is associated with records of achievement processes and their effect on the learning situation. Extrinsic motivation, by contrast, relates to the more conventional allure of the qualification market-place. Although these two goals are not necessarily incompatible, the fact that our evidence documents a trend away from personal recording in favour of the production of interim summative statements suggests that it is the product, rather than the process per se, that may be proving to have more appeal for pupils. At the same time, our evidence that the recording of personal and academic dimensions of achievement are increasingly to be found together in the context of organised school activities, suggests that the centre of gravity in records of achievement is shifting away from the open-ended personal records and diaries which were such an important feature of many pilot schemes when they started. It would appear that sustained personal reflection is not easy to sustain nor particularly motivating for young people.

(b) Individual differences and social divisions

It is important however, to qualify any such assertion with the proviso that pupils cannot be regarded as a homogeneous group in this respect. In records of achievement, as in all other aspects of school life, pupils' responses reflect an immense diversity of personal and group characteristics such as personality, confidence, attitude, achievement level, gender, ethnic background and social class which combine in more or less idiosyncratic ways to colour their response to, and ability to profit from, such opportunities.

There are some indications in our report for example, that girls respond more enthusiastically to, and cope better with, the opportunities offered by records of achievement than boys. As might be anticipated, their verbal skills seem to make girls more at ease with the writing involved and they seem more ready to understand and engage in personal reflection. By contrast, there would seem to be some indications that boys have a different, more external, sense of 'audience' and may therefore cope better with the summative 'reporting' aspect of records of achievement.

Equally there appear to be differences in the response of higher and lower attaining pupils to records of achievement. The former appear sometimes to be unconvinced of the value of records of achievement but are prepared to undertake what is required in the conscientious fashion that characterises their response to school generally, often producing very impressive results. Lower attainers, by contrast, who may be hampered by poor writing skills, may produce a disappointing record that gives them little sense of achievement. In relation to comment banks they may be able to achieve relatively fewer statements of achievement.

Thus pupils, like teachers, need understanding and expertise in the processes of records of achievement so that their ability to profit from the opportunities offered is not limited to that directly associated with the procedures of records of achievement per se. However much the principle of only recording **achievement** is adhered to in interim and final summary documents, there will inevitably be differences in the content and quality of what is produced which may tend to reinforce the hierarchy which bedevils every other approach to assessment, reporting and certification in schools. Thus the summative document's potential power to motivate may be correspondingly weakened for many pupils. If, however, pupils, like teachers, can be encouraged to understand and operate the **principles** of records of achievement this may well help them to both feel more positive about the opportunities such novel approaches to learning offer them and to be more able to make effective use of these opportunities.

Thus, for example, our evidence suggests that although self-assessment is the key to the profiling process in some schools, pupils' efforts to apply it are beset by a variety of problems. These include those of superficiality where the assessments have little significance or meaning; unrealistically low ratings because pupils do not wish to seem conceited and the persistence of norm-referenced statements of attainment — 1.2.1 c(i). Other evidence in Part Two emphasises pupils' need to master the language of assessment through which they can articulate their achievements in order to engage in discussion with teachers.

As we point out in Part One of this report, the substantial literature on teacher-pupil interaction in general and on the effect of pupil characteristics such as gender and ethnic background in particular, finds a resonance in the micro-level data we have collected. There would therefore appear to be a need for more sustained interactionist studies of individual differences in the teacher-pupil relations associated with records of achievement and for the lessons of such research to be used to inform in-service work with both **teachers and pupils**. Only when we know more about the language aspects of records of achievement; the stereotypical expectations on the part of pupils, teachers and parents that may limit pupils' horizons; the unequal power relations that may inhibit pupils' capacity to use to their full advantage the opportunities that reviewing provides; and the cultural assumptions that may significantly demotivate pupils whose values are different to those of teachers and schools, can we be **confident** that the positive effects of records of achievement which are already being documented, apply to more than a minority of pupils.

(c) Age differences

Age represents another dimension to the issue of pupil response to record of achievement opportunities. Our evidence suggests that age can be an important variable in the success or otherwise of records of achievement. Although it is not clear from our data that there is any one year-group that can be identified as optimal for initiating records of achievement, three other points do emerge clearly.

The first is the danger of launching a records of achievement scheme across several or all year groups at once since this stretches available expertise and enthusiasm to an extent where maintaining a sufficient level of understanding and expertise becomes highly problematic. Under pressure of time and lacking the necessary time to evolve procedures, the records of achievement initiative is very likely to be a 'bolt-on' burden.

The second point concerns primary schools and the emphasis that schools appear to have placed on disseminating the **principles** of records of achievement to younger pupils rather than, as might have been expected, a concern for the provision of a primary 'record of achievement' to serve as base-line data for secondary school entrants.

The third point is the need for schools to consider in what ways the procedural expression of records of achievement principles might need to alter for different age-groups within the school. The need to maintain a degree of freshness over time plus the changing focus of concern as pupils mature and become increasingly concerned with the world of work, suggest that records of achievement procedures may need to alter quite substantially during the process of secondary schooling.

(d) Teacher-pupil relations

There is clear evidence that the introduction of records of achievement has led, in many instances, to improved teacher-pupil relations. Not only has it typically provided opportunities for pupils and teachers — especially pastoral tutors and form teachers — to get to know each other better following the introduction of personal reviews, it has also helped to encourage learning situations that are more personally valuing and more personally diagnostic for pupils. Inherent in the expression of personal value is the teacher's recognition of pupils' rights and responsibilities in the learning process by providing for a collaborative approach to curriculum, pedagogy and assessment. Where this has happened the impact has been very positive. At the same time, it is clear that where subject teachers in particular provide opportunities to review progress on a one-to-one basis and encourage pupils to set academic and personal targets in the light of needs that they themselves have been helped to identify, pupils' attitudes to learning are likely to be significantly more positive.

1.3.2
Changes in pupils' academic and personal skills

Identifying changes in these respects is even more problematic than those concerning pupils' attitudes. Apart from the impossibility of separating the effect of cognitive and affective factors in this respect, the range of changes possible make any judgement of cause and effect necessarily confined to the impressionistic. Overall we can say that our evidence documents many teachers who have been favourably impressed by changes in both attitude and achievement among their pupils which they attribute to records of achievement. Equally some of the pupils we studied felt the provision of clearer objectives, personal targets and more supportive relationships with teachers had enabled them to make more progress than would otherwise have been the case. Pupils' personal recording and reviews have also helped teachers to become more aware in many cases of pupils' hitherto unknown personal achievements and this has enabled them to help pupils build on their strengths.

The lack of a suitable 'control' group makes it unlikely that a formal evaluation of the impact of records of achievement on pupil outcomes will ever be possible. However, evidence from similar initiatives where formal research is possible because of more clearly delineated boundaries to the target group confirms that improvements in pupils' learning and satisfaction are likely where such elements are built into the learning process.[1]

1.4
Purpose Three: Changes in curriculum, teaching and school organization associated with records of achievement

The fourth of the policy statement's expressed purposes for records of achievement concerns their impact on the school as a whole and the provision being made within the institution to foster pupil development on a broad range of fronts. In this respect it is relevant to distinguish between the effect of record of achievement activities on individual teachers and their effects on the school as a whole.

1.4.1
Impact on teachers

Our evidence supports the conclusion that teachers' morale is improved to the extent that they perceive a positive response in pupils as a result of record of achievement activities. It also appears to be the case that teachers are defining aims and objectives more clearly for themselves and for pupils and that in some cases at least, there is an improved 'team-spirit' within and across departments in a school. In particular many teachers have been able to use the combined advent of GCSE and RoA as the basis for significant changes in their teaching and assessment techniques to support and encourage the development of pupils' diverse talents and skills. The picture that emerges is of a developmental continuum, at one end of which are teachers who remain relatively untouched by such new developments or who are torn by their conflicting demands. Other teachers are at various points along the continuum with some, at the far end, who

---

1. Batten's 1988 study of the impact of the Victoria STC alternative school certificate programme in Australia for example, provides clear evidence in this respect.

have succeeded in fully operationalising a pedagogy which has clear objectives, individual targets and offers pupils a full partnership in the learning process.

One of the most notable trends in the organization of records of achievement within schools is the progressive blurring of boundaries between what were previously more likely to be regarded as distinct elements of records of achievement.

1.4.2
Impact on schools
(a) The blurring of boundaries

Academic/pastoral. Firstly, the well-established boundary between 'academic' and 'pastoral' is being increasingly overtaken by the recognition that knowledge, skills, attitudes and qualities are all interrelated and cannot be divided into different domains. Several other points follow from this including:

(i) The tendency for schemes to abandon 'personal recording' except in its most basic form of keeping a 'log' of activities and for recording therefore to become more explicitly focused on those achievements that can be demonstrated through the school's educational programme. This in turn exerts a pressure on schools to make more explicit provision for pupils to develop a comprehensive range of skills, qualities and experiences and at the same time, reduces the 'confidentiality' problem associated with personal recording which is not linked to specific school-based targets.

(ii) A further aspect of this change which appears to be both cause and effect is the progressive incorporation of records of achievement into the 'public' reporting system of the school and with it, the introduction of 'interim summative statements' which, in some cases have replaced traditional school reports to parents once or twice a year.

(iii) Cross-curricular recording, always one of the most characteristic and novel aspects of record of achievement activity, represents the most long-standing and explicit attempt to blur these boundaries. However, though our evidence in Part One points to a variety of ways of conceptualizing such achievements, there is still no very clear indication of how such recording can most fruitfully be pursued.

Formative/summative. A second blurring of boundaries concerns the hitherto well recognised distinction between the 'formative' and 'summative' puposes of records of achievement. Our data suggest (Part One section 1.3.3) that this is essentially a distinction in terms of the record's **purpose** rather than of its **content**, at least with regard to interim summative documents. The final document still embodies some distinctive characteristics which set it apart, notably, its exclusively positive content and its explicit ownership by the pupil concerned.

'Ownership' is a much used concept in relation to records of achievement and therefore requires careful definition. It can mean pupils' 'ownership' of the document and/or its supporting procedures. Additionally it may refer to the school's 'ownership' of the record of achievement processes within an LEA scheme. In both cases it tends to reflect the assumption that the **control** involved in 'ownership' will lead to a greater commitment on the part of the 'owners'. Yet another distinction needs to be made however, between literal ownership — as for example, in the case of pupils 'owning' their summary records of achievement, and the more metaphorical notions of 'ownership' as control outlined above.

(b) Ownership

At the present time, it is possible to detect a trend for 'ownership' in the first sense to mean ownership of the processes involved in records of achievement, the school's commitment to discussion and agreement which characterises information recorded for the records of achievement, rather than absolute ownership of documents. Although this is a clear move away from the more robust notions of ownership embodied in the original policy statement it has the explicit advantage of allowing records of achievement to become the core of a school's accountability and reporting procedures rather than a bolt-on optional extra.

In the context of impending National Curriculum assessments this dilemma presents itself in a particularly acute form (see RANSC evidence to TGAT). If pupils are not to be able to negotiate whether or not such information goes on the reports to parents, the danger of the records of achievement procedure losing the benefits that 'ownership' represents for pupils are therefore considerable. Many teachers are explicitly aware of the dangers involved in records of achievement being 'highjacked' in this way. If however, the records of achievement process does not incorporate such assessments it may well become marginalised and redundant. How this dilemma may be resolved will be an important issue for the Records of Achievement National Steering Committee to address.

Questions concerning who controls what in the records of achievement process also highlight the important issue of individuality in records of achievement. Our evidence suggests (Part One section 1.2.3) that the nature of teacher-pupil talk in the review process, for example, is significantly affected by the type of records of achievement system concerned. Where categories of performance are pre-specified as for example, in comment banks, the discussion tends to be about the meaning of the given criteria and agreeing which ones have been achieved. In more 'open-ended' schemes pupils have more opportunity to reflect on their individual achievements and needs. This situation offers a greater possibility of reinforcing their sense of ownership. At the final summative stage particularly, the 'wholeness, uniqueness and context of an individual's achievement' (section1.3.4 (d) of Part One) is perhaps even more important.

There is thus clearly a need to consider closely the interrelatedness of issues of ownership, targetting and assessment. Our evidence suggests that teacher-pupil discussion is enhanced by an explicit purpose or focus, but that it can be spontaneous and can meet a wide range of purposes in a number of different situations. The conclusion that would appear to follow from this is that pupils' must have a clear and explicit right to confine their involvement in records of achievement processes which are not and cannot be entirely within their own control, to aspects of the educational process itself. They must be helped to acquire the skills involved in self-assessment and constructive discussion and in the making of judicious judgements about what may usefully be recorded. As such they will become active and competent **partners** in the 'main-line' assessment process rather than controllers of a 'branch line'.

(c) Curriculum change

Records of achievement are explicitly intended to make fundamental demands on teachers and schools in terms of their curriculum and organization (Purpose Three). It follows that all but the most exceptional schools will expect a degree of change to be associated with the introduction of records of achievement. Our across-site analysis confirms that this is indeed so; that particularly where the starting point for records of achievement is a fundamental re-examination of curriculum, teaching and recording throughout the school, rather than simply a change in the way achievement is recorded, there have been visible changes in approach.

Related to this has been the move towards devising criterion-referenced frameworks of assessment that are realistic and workable. Many externally imposed criterion-referenced frameworks such as the national and subject criteria for GCSE, although they were recognised as expressing very comprehensively the elements of the subject in question, proved unworkable in practice if assessment was not to take up teaching time unreasonably. There is a far greater tendency to organise learning into shorter-term units of work, with stated targets at the beginning and regular recording intervals on specified criteria.

'We observed teachers taking more time to explain the objectives of units of work to pupils, pacing their work more slowly but frequently checking progress with pupils. In order to accommodate such changes, syllabuses were scrutinised to see whether content might be cut so that what was learned might be what was really essential and that it might be learned more thoroughly. In our view such changes are very worthwhile and should be encouraged.

We also observed that in a number of schools where regular teacher-pupil discussion in classroom contexts was regarded as an important part of recording processes in subject areas, the implications for classroom management were similarly positive but far-reaching. For example greater emphasis was placed on individualised or group tasks, independent learning and/or team-teaching. These were regarded as important strategies not only in their own right but also as the only practical way to accommodate more frequent teacher-pupil discussion in lesson time.' (Part One 2.4.3)

We also found that, in reporting, teachers were increasingly citing evidence for their assessments, making these more comprehensive; increasingly emphasising the positive and providing for such records to be negotiated and discussed with the pupils. Most significant of all perhaps was the trend towards seeing the curriculum as **what the pupil takes away rather than what the school provides — in terms of what is learned rather than what is taught**. (Part Two)

(d) Time

This kind of fulfilment of the original vision of the records of achievement policy statement raises two related issues. One is the clear indication that where records of achievement become organically integrated with the teaching and learning philosophy and practice of the school the problem of time to make and discuss 'profile' assessments is substantially reduced. Team-teaching, more pupil-directed learning and group-work can free teachers to spend more time with individual pupils and they illustrate some of the ways in which teachers have managed to make records of achievement a positive support to their teaching through providing for a more varied approach to pedagogy rather than regarding them simply as a time-consuming chore. Secondly we have found that where this integration has been achieved, it is also more likely that explicit provision will be made to allow pupils to demonstrate the achievements, skills or qualities in question.

1.5
Overview

In this section we have tried to present a brief summary of the ways in which the aims of the 1984 policy statement appear to being met within the pilot schemes. We have also tried to clarify how far the four identified purposes are dependent on each other. It would appear that the recognition of achievement in records and reports (Purposes One and Four) is a necessary, but not a sufficient condition for the realisation of the core principle of improving learning (Purpose Two). For this latter aim to be fulfilled, process as well as product criteria must be met. If schools and teachers are not changed by records of achievement (Purpose Three), pupil attitudes are also unlikely to be intrinsically changed.

At the heart of this analysis are questions concerning what the outcomes of schooling should be for young people. There would appear to be agreement that all young people should be provided with a learning environment which allows them to make maximum use of their potential and with a record of their achievements which gives them the best possible chance of being happy and satisfied in their future lives. The key to the realisation of both these goals appears to be the incorporation of processes into the life of the school which have the power to transform the negative messages that many pupils have experienced in the past. The provision of records and reports by itself cannot achieve this. Rather it is clear that such documentation is the catalyst which can bring about a more fundamental change in understanding on the part of both pupils and teachers about the ingredients necessary for all pupils to experience pleasure and fulfilment in their learning. The evidence we have reported, particularly in Part One of this report makes it clear that relatively few schools have as yet fully worked through the implications of this philosophy. That is to say, few as yet have fulfilled all four of the record of achievement purposes. What they have done, however, through their efforts — and this is very important — is to clarify the issues involved. Furthermore they have provided palpable and powerful testimony that the rationale of the initiative itself is educationally sound.

The first section of this final part of our evaluation report has set out in broad terms how far our evaluation evidence supports the validity of the rationale for records of achievement as expressed in the government policy statement that informed and guided the pilot schemes. These conclusions have been stated with little or no reference to the conditions that might need to prevail for such outcomes to be achieved. Thus we now turn to this second, vital area in an attempt to conceptualise what the evidence we have can tell us about how to achieve such successful outcomes. Our emphasis, in this respect, will be very much on school-level variables. This, in part, is a reflection of the fact that the bulk of our evaluation work was couched at this level. It is also a reflection, however, of those schools' own primary concern at present, with institutional variables and their correspondingly reduced concern with external issues such as validation and accreditation. Although our evaluation included a separate strand explicitly concerned with LEA and supra-LEA consortium policy issues, our data in this respect record a highly fluid or unrepresentative situation that leads us to be wary of drawing premature conclusions. Many of these issues need and are receiving further study but some preliminary insights that we do have are reported here.

First and most fundamentally, we want to suggest that the impact of any records of achievement scheme will be in direct proportion to the **sense of ownership** of the procedures that is felt by the teachers and pupils involved. The capacity of a record of achievement initiative to meet its avowed purposes would also appear to be a direct function of the extent to which it has become 'an active principle of the school's working life' — a dimension which we refer to as 'penetration'. It is these two principles of 'ownership' and 'penetration' which we have come to regard as the defining principles of a successful record of achievement scheme. Notwithstanding the existence of national or scheme-wide guidelines which may be necessary for the purposes of quality-assurance and credibility, it is clear from our data that fulfilment of all four purposes of the Policy Statement depend upon schools having the opportunity to evolve ways of meeting RoA criteria which suit their particular institutional needs and traditions. To the extent that they are involved in devising such procedures, **teachers and pupils** are likely to feel a corresponding sense of ownership which is important for sustaining support for records of achievement. It is also critical if records of achievement are to become an organic part of school life, as an informing principle of the whole variety of activities rather than an identifiable innovation. Both 'ownership' and 'penetration' are in turn dependent on four further elements which can be broadly labelled as 1) understanding and expertise 2) credibility 3) practicability 4) commitment — as set out in the following diagram:

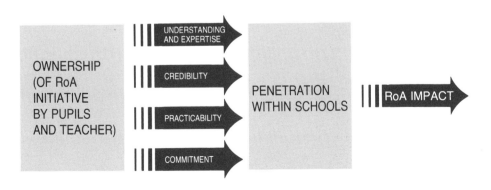

A model of the elements required for the successful development of a records of achievement scheme.

Clearly such outcomes will only be achieved if pupils and teachers understand the principles involved and are able to devise suitable procedures to meet them. INSET is obviously very important in this respect and this is recognised by virtually all schools and schemes. As the relevant sections in the report set out, such INSET needs to cover a wide range of areas including the skills needed to analyse curriculum content so that it can be broken down into assessable targets; assessment approaches and techniques; summarising skills; discussion and review skills, and approaches to the effective introduction and management of a records of achievement scheme. Form tutors in particular need training in the skills needed to work with pupils on the selection and summarising of information and often in the actual writing of summaries themselves. Some of the training needed may well be provided for in INSET associated with National Curriculum assessments.

In addition to these demands which records of achievement are making on teachers are the skills needed to actually conduct reviews successfully. As we point out in Part One, our evidence tends to support the expectation that it will be difficult for teachers in such situations to change traditional patterns of classroom discourse at a stroke.

'It would be surprising indeed if these powerful patterns were miraculously absent from record of achievement discussions, simply because the nature of the task was slightly different from those to which teachers and pupils were accustomed. One very interesting piece of evidence emerging indirectly from our work is that teachers to whom we have sent transcripts of tape recordings we have made of their discussions with pupils have, almost without exception, been quite horrified to see just how much they talk and how little their pupils talk, how controlling their talk has been and how submissive and compliant their pupils' talk has been, and how quick they are to 'fill up silences''. (Part One Section 1.2.3(b))

At its most general level, this analysis highlights the critical need for understanding and expertise in records of achievement on the part of teachers. Like pupils they must be helped to acquire a full understanding of the **principles** that underpin the specific **processes** that the school has chosen to institute as the vehicle for records of achievement. Only where this is the case will the critical dimensions of ownership and penetration apply. That this is so is clearly evident from our data which strongly reinforce the point that it is where teachers are committed to records of achievement philosophy and are able to use it organically in their teaching and/or tutorial work, that pupils in their turn respond positively. This philosophy is cogently expressed by a Lancashire teacher:

'I have always thought that records of achievement involved changing hearts and minds. It is about fundamental differences in the way in which we teach and children learn. It is not bolt-on; it is an over-arching membrane which gives strength and purpose to what we do, and underpins our very philosophy.'
(Lancashire final report to RANSC)

The evidence suggests that national, LEA, consortium and school-based INSET all have a useful role to play provided that in every case it is closely connected with practice. Existing research evidence[1] suggests that acceptance of the principles behind change depends on teachers being convinced through their own experience of its desirability. Our data also point clearly to the vital interrelation between theory and practice in records of achievement development work where, on the one hand, opportunity to examine the rationale of records of achievement is a vital step in the development of commitment and good practice among teachers and, on the other, teachers often remain

---

[1] See, for example, Doyle, W. and Ponder, C.A. *The Practicability Ethic in Teacher Decision-making*, N. Texas State University (mimeograph) (1976).

sceptical until convinced by their own experience. Thus INSET must help teachers to develop understanding of, and commitment to, the rationale of records of achievement and also to equip them with the specific skills they feel **they** need to apply such principles in the idiosyncratic context of individual classrooms and schools. In some cases this has meant simply giving groups of teachers within a school the opportunity to meet together and work through issues in their own way rather than providing any specific external input. Informal support networks, organized by individual schools and teachers themselves to provide for mutual exploration and support through the exchanges of experiences, have provided a very useful framework for INSET delivery. The involvement of many more schools in such mutual support arrangements is to be encouraged.

2.1.2
Pupils

The need for pupils also to be given the opportunity to develop their understanding and expertise in relation to record of achievement processes is much less commonly recognised. Although lip-service is often paid to this need, we have encountered few examples where formal provision has been made for training in record of achievement techniques such as self-assessment and negotiation skills. More typically subject teachers and reviewing tutors appear to anticipate the development of the appropriate insights and skills as familiarity with the procedures increases. Our evidence suggests that more explicit efforts to provide for such competencies among pupils would be a significant step towards making record of achievement procedures more effective and by the same token, considerably facilitate the task of teachers who are often faced with the necessity of breaking down deeply-rooted assumptions held by pupils about the respective role of teacher and pupil in the classroom.

2.1.3
Other partners

Understanding and expertise are also important requirements for the other partners in the records of achievement process. As the NFER study on employers' reactions has already revealed, employers need to be encouraged to ask for records of achievement if they are to be fully used. Some schemes have taken this very seriously. Wigan, for example, has provided substantial INSET in the form of, for example, residential weekends for parents, governors, employers and community representatives such as LEA elected members.

Although partly informed by a concern with credibility, it is also clear that many schools and schemes recognise the key formative role that such groups can potentially play in the record of achievement process. In the across-site analysis, for example (section 1.3.3.) we refer to the need suggested by our data, for parents to be one element of a tripartite partnership so that, rather than providing them with a potentially misleading and frustrating interim summative record which is entirely positive, they too can be drawn into the formative process that addresses strengths and needs and so be helped to acquire the understanding and expertise that will enable them to add further to the efforts of pupils and teachers.

It follows from this that organizations receiving young people on work-experience for example or members of the community involved with validating activities for the records of achievement must also be helped to develop an active understanding and acceptance of record of achievement principles if these aspects of record of achievement activity are not to be mere bureaucratic exercises. Although greater credibility for the summative record of achievement will undoubtedly help in this respect it also, paradoxically, carries an inherent danger of over-emphasising the extrinsic motivation associated with the products of the record of achievement process which may lead to a corresponding disregard for the more fundamental potential of the process itself.

A related problem is the difficulty of getting all sections of the community involved. Like pupils and teachers, parents, employers and other users cannot be regarded as homogeneous groups. The differences in perspective and practice between large and small employers, for example, are paralleled by equally significant differences among parents from a wide range of socio-economic circumstances. Thus care will be needed that assumptions based on values and practices that are specific to a particular group are not so built into the record of achievement process that substantial sections of the community are out of sympathy with them or unable to participate in them. If this happens it will erode any potential for a real partnership between home, school and the world of work to develop.

2.2
Credibility
2.2.1
Credibility in the context
of national assessements

The need for all those associated with records of achievement to understand and support the goals of the initiative, suggests that the issue of credibility is a much more profound one than simply that concerning the currency of the record of achievement for school-leavers. At its most profound it will reflect the extent to which public opinion can be convinced that the novel approaches to teaching, assessment and certification associated with records of achievement are at least equally, if not more successful in bringing about learning and raising standards than more conventional practices in these domains. With the steadily increasing emphasis on the 'evaluative' function of assessment as a mechanism of accountability to parents which has characterised recent initiatives in national assessment policy, pressures for the use of more traditional forms of assess-ment, and its expression in more traditional numerical forms may well increase. Given that the institution of national assessments at ages 7, 11, 14 and 16 may, in particular, encourage consumers to regard these results as the hall-mark of pupil and school achievement, teachers will be called upon to defend the distinctive philosophy and practices of records of achievement since the latter have themselves been shown to have considerable potential for raising standards. If not there is a danger that records of achievement could be reduced to being simply the vehicle which provides for the recording of such results, as the wording of the 1988 TGAT report[1] indicates:

'In particular the moderation and national testing required by the national curriculum could furnish a basis for accreditation and validiation for some components of the records of achievement. These in turn provide a means of recording achievement through the secondary years (indeed starting at the end of the primary phase), including that related to personal and social development; and all the more valuable if pupils are drawn formally into discussion with teachers and parents about their progress. We therefore recommend the use of Records of Achievements as a vehicle for recording progress and achievement within the national assessment system.'

The issue here is how far records of achievement can become integrated as a pivotal feature of school life if they are simply used as the basis for discussion about, or the reporting of, assessments. The need which our evidence has identified for ownership and penetration suggests that the successful realisation of DES policy in this respect requires schools and teachers to resist any alteration in existing record of achievement priorities. In this task they may well find themselves aided by the parallel commitment to positive, individualized and carefully-targeted assessment which will inform the national assessment.

Alternatively, schools and schemes, local and central government may find themselves faced with a stark choice between two different assessment purposes – 'assessment for curriculum' versus 'assessment for communication.' It may be that records of achievement will succeed so well in their own terms that they fulfil both purposes and gain support from teachers and users alike for the visible difference they have made in pupils' attitudes and achievements. If so, they will also have enhanced the school's ability to meet the more conventional criteria of pupil success. Clearly nothing can yet be said with any evidential base

---

[1] TGAT report 1988.

about the impact of the momentous changes about to affect the education system. Our evidence does however help us to formulate some of the important questions in this respect as far as records of achievement are concerned and to stress the importance of RANSC and others addressing them at the earliest opportunity. These include:

(i) How can schools work to prevent records of achievement being reduced simply to an elaborate recording process?

(ii) How can schools ensure that the necessary changes in teacher-pupil-parent relations, curriculum framework and teaching styles that our data have shown to be crucial to the success of record of achievement policy objectives, are not inhibited by other policy developments?

(iii) How can schools ensure that the credibility of records of achievement continue to concern the procedures associated with it, rather than just the summary record of achievement and its perceived utility to users as a supplement to more conventional forms of certification?

(iv) How can schools protect those record of achievement processes that our evidence shows to be critical in changing pupils attitudes and capabilities?

(v) How can the situation be avoided in which the record of achievement becomes just another demotivating influence for some pupils, a school-imposed and controlled procedure that serves only to record their relative lack of progress compared to other pupils – a procedure which engenders no sense of ownership, either of process or product?

(vi) Above all, how can schools avoid implementing a scheme which makes pupils feel they are constantly being judged, measured and evaluated in every aspect of their life, a sense of surveillance and control that would be quite alien to the intentions of records of achievement?

Our evidence already provides examples of pupils being 'guided' to include some achievements and experiences rather than others. It is not hard to see how readily the pervasive influence of record of achievement procedures, if they become linked with provision for institutional evaluation and 'performance indicators', could become a powerful form of individual surveillance and of social control generally encouraging pupils to conform to a particular set of social norms and devaluing cultural variations in experience and attitude. Some evidence for these arguments, which have been expressed in more detail elsewhere already exists[1] and is set out in the discussion of power relations with respect to records of achievement in Part Two of this report.

Impending developments in the educational context will inevitably be very significant for the future of records of achievement and it will be vital to take this into account in seeking to build upon the conclusions being drawn from the pilot scheme experience.

2.2.2
Procedures for quality assurance

Absorbed as they have tended to be in designing and introducing the actual procedures to support recording achievement, some schools have so far tended to show relatively little interest in procedures for quality assurance. For the reasons given above, this lack of concern is unlikely to persist. Those committed to the success of records of achievement initiatives will be anxious to employ every means possible to raise the status of school-leaving statements

---

[1] See, for example, Hargreaves, A. 'Record Breakers' in Broadfoot, P. (Ed) *Profiles and Records of Achievement: a review of issues and practice*, Cassell, London (1986) and Broadfoot, P. 'The significance of contemporary contradiction in educational assessment policies in England and Wales' in Stake, R. *Effects of Changes in Assessment Policy*, C.I.R.C.E. University of Illinois.

among potential users. The history of similar proposals in this country to provide school leavers with a school testimony to their achievements, reveals how bedevilled such initiatives have typically been by their failure to achieve credibility among potential users. In the face of the elaborate modernisation procedures operated by examination boards to ensure reliability and comparability in public examinations, other kinds of certification lacking such controls have inspired little confidence. Thus, whilst urging the need for a major publicity campaign at national level, many schools and schemes feel that the imprimatur of an examination board, added to that of the LEA or record of achievement scheme, will inspire greater confidence among users.

Traditional product-based approaches to moderation however are clearly not appropriate for records of achievement where the products are so highly individualised and it is the processes supporting the production of the final record, as much as the statement itself, which must be monitored. Thus the quality assurance procedures so far devised in relation to records of achievement have tended to draw on practices developed in further and higher education in relation to the 'validation' of courses of study and the 'accreditation' of institutions to issue particular awards in the context of records of achievement development. These two principles are often used almost interchangeably and certainly no generally agreed definitions of them as yet exist (See Part One section 1.6.). The tendency to conflate these two traditionally separate procedures in the context of records of achievement in part at least reflects the inextricable relationship in records of achievement practice itself between the nature of the learning experiences offered and the associated procedures for assessment. As assessment has increasingly become a pedagogical, as well as a judgement device, record of achievement schemes have found it necessary to develop novel quality assurance techniques which are typically still in the process of evolution and do not readily fit any of the traditional approaches. Generally speaking, however, these procedures currently involve the accreditation of individual schools by an outside body when that body is satisfied that certain key process criteria, which have been agreed as central to the successful operation of a record of achievement scheme, have been satisfactorily implemented in that institution.

Although our evidence is relatively thin in this area, it is sufficient to raise a number of questions about this approach to accreditation. For example, is it the quality of the processes concerned or only their existence which will be the basis for judgement? What is the most cost-effective approach to such arrangements: inter-school visitation with all the dangers this has for professional rivalry or the use of the limited resource of LEA advisers and inspectors' time? How can such accreditation be arranged to provide for the most beneficial effects on schools: through schools self-report perhaps and by encouraging schools themselves to identify what the criteria for effectiveness might be?

Other schemes emphasise consumer satisfaction as the key to successful quality control. If 'the proof of the pudding is in the eating' can validation of school processes be left to the market-place? One argument against this is that the criteria of professional educationists about what a record of achievement scheme should be seeking to achieve may not be shared by the outside world with its much more utilitarian concerns.

The issue of 'whose criteria' is likely to be further complicated by the advent of national assessments the moderation of which 'could furnish a basis for validation and accreditation for some components of the records of achievement' (TGAT report, paragraph 162). Although this is currently the case with external examination and graded assessment results included on the record of achievement, the addition of national assessment information may well serve to strengthen the emphasis on product moderation in records of achievement rather than its novel attempts to introduce the validation of educational processes. One outcome of such a development might be to hasten the degeneration of records of achievement schemes into an alternative or supplementary form of certification rather than a coherent approach to the educational process itself. Alternatively, the inclusion of such information on the record of achievement might lead to the latter itself becoming the main form of certification, and the associated procedures the principle focus for external monitoring.

A third option in providing for quality control is that of school-based validation and accreditation boards. In Dorset, for example, despite the putative introduction of the Southern Partnership for the Accreditation of Records of Achievement (SPARA) to provide a broad regional currency for records of achievement, the primary concern has been to establish both Validation and Accreditation Boards. These are composed of local community and school representatives which work closely with individual schools to support and guide their efforts to institute records of achievement arrangements which are 'an active principle of the school's working life' developed in each case by the teachers concerned to provide procedures which are fully 'owned' by that school. This approach to providing 'credibility' for records of achievement has the advantage of encouraging the development of understanding and expertise among the school's local partners. It also serves to reflect back its own practice to the school in a more immediate way, so encouraging on-going reflection and development on the part of those involved.

No judgement can yet be made about which approach to quality assurance is likely to prove the most satisfactory in the longer term. Indeed this will be one of the most important and difficult issues that the National Steering Committee will have to address. If, as we have argued above, the power of records of achievement to sustain its current momentum towards fulfilling the goals of the policy statement, depends critically on public perception of its impact on pupils; and if, as we have also argued, this latter depends on the extent to which records of achievement are organic to the life of the school, we may speculate that the third approach to validation and accreditation described above perhaps offers the best compromise in terms of both credibility and utility.

2.3
Practicability
2.3.1
Context factors

Records of achievement have become a major plank of national policy at a time of momentous change for education in this country. Firstly the initiation of the records of achievement pilot schemes coincided with the introduction of a major new examination, the GCSE, which made novel and unprecedented demands on schools and teachers. Our evidence testifies to GCSE having been both a help and a hindrance to records of achievement development. On the one hand, it helped to familiarise teachers with, and in some cases provided useful training in, criterion-referenced assessment approaches. In many instances too teachers were able to adopt GCSE assessment requirements to meet the needs of their records of achievement procedures. Thus in one sense it can fairly be said that GCSE was a very significant positive factor in supporting the introduction of records of achievement within the pilot schemes.

Set against this must be the demands that the GCSE itself continues to make on the limited time and energy resources of teachers which, together with its necessarily high priority in schools, have combined to ration severely the available resources for records of achievement developments.

If it is further taken into account that the first year of the national records of achievement pilot was affected by teachers' action in schools on an unprecedented scale which was followed by new contractual and training arrangements for teachers and that schools are being bombarded by a range of new curriculum and assessment initiatives, it is clear that the practicability of records of achievement has been tested against a background of the most adverse circumstances.

It is also true that even more momentous changes are still pending with the introduction of the 1988 Education Reform Act provisions. Although at least some schools have already laid the foundations of their records of achievement scheme and have useful experience to offer other schools in this respect as a result of pilot schemes, it nevertheless is the case that in the future the context for records of achievement development will be one in which the resources of time, energy and skill are stretched even more than they have been within existing, specially resourced pilot schemes.

This makes it even more imperative that commitment and ownership on the part of teachers and pupils should characterise the introduction of records of achievement schemes. Not only does this help to minimise the hostility which may result from what is perceived as excessive pressure from external initiatives, it also helps the school to integrate and co-ordinate the demands of these same initiatives along with existing procedures in such a way as to make the fit as efficient and economical as possible between the various needs which have to be met in the particular context of each individual school.

This further suggests the need for an assessment policy on the part of both the school and the LEA which will provide the formal means of making arrangements for such integration. Our evidence suggest that schools have not found such integration easy, nor indeed have they found it easy to make records of achievement an umbrella for a range of different initiatives as many of them intended. It is likely that such difficulties are a reflection of the relatively early stage that such attempts at co-ordination are typically at, but they may also reflect the less than total integration of one or more of the initiatives into the life of the school which makes it difficult for them to be addressed in a coherent way. This in turn may be a reflection of the lack of such a coherent policy at local authority level which our evidence suggests is equally crucial. It is clear that difficulties inherent in the formulation of a school assessment policy reflect the varied and sometimes contradictory purposes of assessment and the differing priorities between these purposes that various sections of the school community may afford. In this respect records of achievement may prove a useful catalyst in encouraging schools to work through these differences to achieve a greater coherence in their assessment approaches.

2.3.2.
Time and resources

A great deal has been said and written about the question of time and resources in relation to records of achievement and our evidence confirms the existence of these worries almost universally. If 'resources' are defined in their broadest sense to include time and space for teachers to undertake records of achievement activities, especially one-to-one reviewing; the time of ancillary staff especially in the crucial task of preparing school-leaving documents; materials and equipment including computers, storage facilities and stationery; time and resources for INSET and even the time of LEA personnel for accreditation where this is applicable, it is clear that, as many of the pilot schemes stress, the successful implementation of records of achievement does have major resource implications. Our evidence suggests that much of the burden of records of achievement is currently falling on form tutors and that this is associated with several other time problems. First, many 'reviews' are being conducted between teachers and pupils who don't know each other and that much precious review time is being taken up with the partners getting to know each other, rather than focusing on the specific task in hand.

Secondly, even where tutors do know their tutees — perhaps because tutor groups have kept the same tutor for a number of years — such reviews are unlikely to be closely focused on targets in individual subjects since the tutor will not be in a position to help the pupil to make specific future plans for each subject. Instead the tutor's role is typically one or both of two more general tasks. One of these is the more pastorally-orientated concern of providing for reflection by the pupil about him or herself as a person across and beyond the formal curriculum; of developing for each pupil a curriculum overview and providing guidance in relation to it. The second task for tutors is typically that of co-ordinating the production of interim and final summative statements. This task may make very substantial demands on tutors not least in terms of the obvious time involved and points to the need for subject teachers to find more time to provide for this kind of reviewing as part of normal classroom routine. Indeed many schools in our study felt that such provision could not be made successfully without enhanced staffing.

It is also true that for many teachers resource issues have a further, less instrumental dimension. The provision of additional resources on the part of the LEA or indeed central government is seen by teachers as a statement of commitment to the initiative which is in turn a strong influence on their motivation to commit scarce resources of time and energy to the record of achievement development work.

Our evidence points also to the need for a re-direction of priorities which will allow records of achievement to have a strong and legitimate claim on available resources. Whether this happens is likely to depend on teachers' enthusiasm for records of achievement and the extent, once again, to which they feel it is **their** creation and something into which they wish to put scarce resources. It also, by contrast, depends on factors quite outside teachers' and schools' control namely the strength and perceived legitimacy of parallel demands on schools to which other policy changes are likely to lead.

2.3.3 Computers

It seems possible that some ways may be found in the future of economising on the time needed for records of achievement — notably through the use of computers — although such attempts appear so far to have been dogged by the mechanistic approach to assessment that has typically characterised them. Where however, appropriate software can be developed and sufficient hardware made available to provide for a pupil-controlled system of recording and storage, computers would appear to have much more potential. That is to say, it may be that it is in their word-processing mode, rather than in their data-processing mode that computers will have the most potential in both an educational and a pragmatic sense, although this does not rule out the possibility of combining the two as section 1.3.5 of Part One suggests.

In sum then, it may be concluded that to the extent that additional resources can be found to provide for review time and meetings, materials and equipment, INSET, ancillary support and accreditation, this will immensely facilitate the records of achievement process. But although such resources may well reflect, they cannot, in any sense, compensate for the commitment of those involved which is, we argue, the ultimate key to success or failure.

2.4 Commitment
2.4.1 Within the educational system

In Parts One and Two of this report we identify various dimensions of the issue of commitment which have emerged as important from our analysis. These include the explicit commitment of the school's senior management team and its headteacher; that of the LEA and of central government itself. In particular the perceived level of central government commitment, whilst it might not affect the enthusiasm of teachers themselves, would be likely to have an effect on the relative priority given to records of achievement in the decision-making process.

2.4.2 Co-ordinators

It is also clear that the school co-ordinator's role is a critical one. His or her status, time allocation, skill and above all, personal qualities have been consistently identified as of major significance in the potential success of a records of achievement scheme. The finding that it is sometimes difficult to give sufficient free time to someone of appropriate seniority and talent who is likely to have many other calls upon them for that very reason, reinforces once again the need for records of achievement to be organically integrated into the life of the school. The suggestion, for example, which we refer to in Part Two that schools may need to increase the number of deputy heads to provide adequately for this role is a good example of the kind of organizational change which the policy statement explicitly looks for in the implementation of records of achievement. This will however, be very difficult to achieve without whole-hearted understanding and commitment to records of achievement within the school and in many cases, extra resources.

2.4.3
Working groups   A similar point emerges from the evidence presented in 2.1.2 of the across-site analysis where it is suggested that the schools studied had many different reasons for opting into a records of achievement pilot scheme which included the desire to change the school, the reporting system, teachers' notions of achievement, and the motivation and self-esteem of pupils and teacher-pupil relations. The evidence suggests that:

'the chief importance (of these starting points) has perhaps been in the way that they have determined the overall character of the scheme and influenced the attitudes of staff. Expressing the purposes of the project in terms which incorporate the aspirations of staff may therefore serve not only to provide a 'selling point', but also to imbue the project with a wider significance which extends naturally from existing priorities.' (Part One section 2.1.2)

At the same time we argue that devolution of responsibility for the development and running of the records of achievement scheme to various working groups within the school was a critical feature in the development of a sense of understanding and ownership among staff.

Probably the aspect of consultation which was most crucial for the active understanding and acceptance of the project among staff was the quality of information which was made available, in terms of the nature of the scheme, the objectives on the part of senior management, the kind and degree of priority which senior management placed upon the project, the role which staff were expected to play and the ways in which the project related to their existing practice and principles. This quality of information often determined whether staff perceived the project as essentially democratic or autocratic, and whether consultation was genuine or cosmetic. (Part One section 2.1.3)

In this brief drawing together of some of the main findings from our evaluation it has been our aim to highlight the major themes which are raised by records of achievement and the factors which appear to determine the success or otherwise of a records of achievement scheme. In particular we have pointed to the twin principles of 'ownership' and 'penetration' which are proposed as an overarching rationale for development work. Schools which are characterised by these elements are likely to be able to move easily between principles and procedures within records of achievement and to recognise the need to do so.

An important feature of the analysis is the way in which what appear at first sight to be issues which are specific to records of achievement development are revealed on closer inspection as more fundamental educational issues. Thus, for example, the existing research on 'self-esteem' and 'locus of control' have much to teach us about pupil attitudes. Learning theory and metacognition research likewise can reveal important principles in the construction of effective learning situations; interactionist and teacher effectiveness research can help us to learn what it is that leads to positive teacher-pupil relations. The extensive literature on curriculum planning and development, the management of innovation and change and on management generally has much to offer that is relevant to records of achievement. These sources need to be explicitly tapped for substantiation and illumination of the emerging insights we have recorded here and it is to be hoped that what we have written will stimulate further studies with this as their explicit aim. Meanwhile it is our task in these closing pages of our report to offer some conclusions to guide policy-making which appear to be suggested by our findings.

We have suggested that the twin principles of 'ownership' and 'penetration' — representing respectively the depth and the breadth of the initiative's dispersal within an institution — are the cardinal principles on which all else depends. We have sought to describe the kind of framework that needs to be provided nationally and locally to support such development and to ensure its quality.

Suffolk has coined the phrase 'the tightrope factor' to describe the flexibility required between a school's unique activities and the framework for records of achievement provided by the LEA. Dorset indentifies five aspects to the LEA's role in this respect: provision of guidance principles; provision for quality control; provision of resources; provision for sharing expertise between schools and provision for co-ordination between initiatives. Beyond the provision of such support, it is clear from the evidence to date that schools must be left to evolve their procedures from the 'bottom-up'.

At national level the same points apply. There is a fine balance to be drawn between providing a context or framework of support and giving schools sufficient flexibility to meet procedural guidelines in their own way. Such an approach is quite contrary to that taken to introduce the GCSE, for example, which like many other assessment innovations involved the drawing up of tightly-defined external guidelines which were specifically intended to bring about assessment-led curriculum change. But the records of achievement initiative is not, essentially, an assessment initiative; it is, in part, a form of curriculum development. At root however it is a pedagogic initiative. This is why it is so important that its introduction is conceived of as an INSET exercise rather than the imposition of a particular set of procedures. The ideal approach would appear to balance the same degree of pressure for change that was associated with GCSE, whilst sustaining the need for institutional autonomy.

Our conclusions depend on a very extensive but still patchy body of evidence which, for a variety of reasons which are set out in the report, still leaves some issues relatively unexamined. For example, we have relatively little to say about the roles of LEAs and consortia, about validation and accreditation, and about the impact of summative documents. These are issues which will handsomely repay further study as their key role in the development of records of achievement on a substantial scale becomes clearer. Meanwhile our primary focus remains the four purposes of the original policy statement, and the fundamental rationale that it sets out for why records of achievement should be introduced. From the evidence provided in our report and the very substantial evaluation data it represents, there can only be one conclusion: records of achievement pose a challenge to schools and teachers that is perhaps unprecedented in formal education. They make novel and substantial demands on time, energy, resources and skill across a wide range of fronts — demands that schools and LEAs have been able to meet to varying degrees. There clearly remains much hard work to be done. Despite all this, there has been no serious challenge to the policy itself and no turning back. Our evidence, like the responses to the National Steering Committee interim report confirms a continuing consensus that records of achievement can raise the standard of pupils' learning by raising their involvement in, their commitment to, and their enjoyment of the educational process. The best evidence of the achievement of the pilot schemes is the achievement of their pupils.

Introductions to school case study reports

BLUEBELL SCHOOL, DORSET
Case study worker: Barry Stierer

School context

Mixed-sex comprehensive secondary school catering for 11 to 16 year olds. Approximately 820 pupils on the roll; staff of 45.

Catchment area was one of mixed housing, ranging from relatively affluent residential surburbs to council estates, although the majority of pupils lived in council housing. Fewer than 5 per cent of the intake came from ethnic minority families. The local area was characterised by economic growth and relatively low unemployment. There were several large local employers.

Overview of the school RoA project

A pilot school participating in the Dorset/SREB RoA pilot scheme.

The main components of the school's scheme were:

(i) Individual departments developed subject profiles with the support and guidance of the school co-ordinator and the Dorset Assessment Team. Each department was given considerable scope for developing its approach to profiling in a manner and at a pace of its own choosing.

(ii) Some tutors introduced new forms of tutorial work, such as regular recording of activities, interests, hobbies and achievements in and out of school ('This is Me').

(iii) A summative RoA was issued to fifth year pupils in the summer of 1987, containing a personal statement which illustrated six personal qualities with 'achievements' in and out of school; a summative 'This is Me' statement written by the pupil; and summative subject profiles produced by some departments.

Main focus of the project in 1985/86 was on year 4, and in particular (i) the development by departments of subject profiles for that year group, and (ii) the introduction by fourth year form tutors of 'This is Me' recording.

In 1986/87 the project spread to include not only the original cohort of pupils (then in year 5) but also the new fourth year and the in-coming first year. A greater emphasis was placed on the work of form tutors in years 1 and 4, who were allocated a weekly tutorial lesson.

PRAISE case study evaluation

12 visits over 18 days were made to the school for the purposes of data collection between November 1985 and November 1987. The main focus for data collection was the experience of staff and pupils in relation to the first cohort of pupils involved in the project (i.e. year 4 in 1985/86 and year 5 in 1986-87). This was pursued through interviews with most heads of department and most form tutors for this year group, as well as with year heads, deputy heads and the school RoA co-ordinator. A special study was made of one tutor group throughout the two years, through observation of lessons and interviews with the pupils. Interviews between a sub-sample of this group of pupils and the school RoA co-ordinator for the purposes of discussing the summative RoA were observed and all parties interviewed. The evaluator attended a number of different meetings held at the school as an observer during the fieldwork period. These included planning meetings attended by various members of staff and various members of the Dorset Assessment Team, and meetings of the school validation board.

The overriding emphasis of the project on the development of formative profiling in subject areas.

The use of 'personal qualities' as headings and prompts for the recording of achievement in and out of school, and the assumption underlying this use, i.e. that **all** pupils demonstrated aspects of these personal qualities and that they should be given the opportunity to reflect upon and record that achievement rather than to judge (or indeed have judged for them) whether or not they had demonstrated the relevant qualities in some absolute sense.

The explicit attention given to the recording of cross-curricular achievement in two ways: looking across the curriculum from the tutorial perspective, and looking at subject-specific achievement as a variant of achievement across the curriculum.

The difficulties in sustaining continuous personal recording ('This is Me'), some of which stemmed from the subject orientation of the system.

The prominence given to pupil self-assessment in formative coursework profiles, and the three types of self-assessment suggested for the purposes of analysis.

The 'double-bind' encountered by many subject teachers developing processes of pupil self-assessment, i.e. minimal comments in response to open-ended prompts, and comments which merely rephrased the teacher's question in response to detailed prompts.

The relative absence of formal teacher-pupil interviews within the system at Bluebell, and the greater emphasis on written communication. Also the implications of this emphasis for pupils with writing difficulties or for pupils who preferred oral media.

The tension within the project between descriptive assessment principles and quantitative assessment principles, and the fact that the very wide range of formative profile documents signified not merely a surface diversity in format and presentation, but also a diversity in underlying assessment principles.

The work of the School Validation Board.

The style of RoA management adopted in the school, which allowed the pace of development and the level of knowledge and commitment to be set by a small vanguard, rather than to wait for consensus before moving to the next step.

The continual change and adaptation to which work within departments was subjected, and the variability of work across departments, was for the most part a planned and valued feature of the project.

The adoption of a policy to send formative subject coursework profiles home to parents as and when they were completed, in lieu of a uniform reporting schedule to parents.

NASTURTIUM TERTIARY COLLEGE, ESSEX
Case study worker: Mary James

College context

Nasturtium Tertiary College was established in 1984 as a result of the reorganisation of the town's existing FE college with local sixth forms. In 1986 it had 1,350 full-time and 3,000 part-time students (2,200 full-time equivalents) and 221 teaching staff. Its student intake was comprehensive and provision included academic, vocational, pre-vocational and post-experience courses at all levels (ranging from GCSE and CPVE to a post-graduate course in journalism). Most students were in the 16 to 19 age range but some were much older and the college also provided elements of TVEI work with 14 to 16 year olds. Less than 5 per cent of students came from ethnic minority groups.

The college occupied a split site in a new town which had above-average unemployment although it was experiencing a period of economic growth. Because of the vocational nature of much of its work and the need to be responsive to the demands of the labour market the links with local industry were strong. The town was the centre for some very large businesses but it also had many diverse small employers.

Overview of the college RoA project

With the appointment of a team of two college co-ordinators in 1985, experimental work was begun with a group of approximately 150 students and their tutors. In 1986 all 16 to 18 year old students on one year full-time courses and the first year of two-year courses were involved. By 1987/88 all 16 to 18 students had some form of RoA.

The objective of the project was to devise a system which could incorporate routine assessment within all courses and provide a uniform basis for regular reporting. Emphasis was placed on developing assessment frameworks based on explicit course objectives, as well as a personal record, and encouraging student self-assessment and recording based on recognition of a need to provide concrete evidence of achievement. Tutor-student discussion and negotiation of records was also an important principle although it proved difficult to convince all staff and students. Two things pushed development forward at a time, half-way through the pilot, when the initiative looked like stalling. One was the development of a student-controlled, user-friendly computer programme for recording; the second was the management decision to make RoAs college-wide policy requiring the practical support of staff at all levels.

PRAISE case study evaluation

Data were collected during eleven college visits from January 1986 to November 1987. A main focus was the experience of one personal tutor group in 1986/87 which drew students from GCE A Level, Sports Foundation, Leisure and Recreation, BTEC Business Studies National Diploma and Intensive Secretarial courses. Personal tutorial and 'record and review' sessions were observed and students and their personal and course tutors were interviewed. Questionnaires were also sent to parents of this close-focus sample of students, but response was very poor. Formal interviews were also conducted with students from a different tutor group in the previous year and with various other members of staff: a Vice Principal, the Head of Student Services, a Dean, a Leader of Division, and the CPVE Co-ordinator. In-service sessions and various planning meetings were also observed and relevant documents collected.

Main issues carried forward for across-site analysis

Four inter-related themes were of particular interest:

First, was an aspiration to develop a system which gave students a large measure of control over assessment and recording in line with the desire of senior management to create an ethos which was responsive to student need and encouraged negotiated learning.

Secondly was an explicit intention to create an RoA system which would act as an umbrella for other forms of assessment, recording and reporting within the college (e.g. CPVE, GCSE, GCE, BTEC, CGLI). The attempt demonstrated that this was possible insofar as most new assessment initiatives were moving in the same direction i.e. towards modularised learning, curriculum objectives, criterion-referenced assessment and profiling. The additional aspects that the RoA project introduced were student self-assessment and recording, tutor-student negotiation, and an explicit requirement to provide prose statements of achievement supported by concrete evidence.

Thirdly was the unanticipated motivating effect of a user-friendly computer software package which was not dependent on any comment bank but which invited students to summarise their achievements, in prose with appropriate evidence, in relation to a framework of criteria or headings designed by their tutors. A number of print-out options enabled the production of formative records, reports to parents and summative documents.

Fourthly were management issues concerning the difficulty of gaining the support of those groups on whom the survival of the project post-pilot was most dependent i.e. the Deans of Study, and the perceived need to establish a college-wide policy on records of achievement which made implementation a requirement. With respect to this last, the tensions and interactions between 'bottom-up' and 'top-down' approaches to innovation and change were well illustrated.

POPPY COUNTY HIGH SCHOOL, ESSEX
Case study worker: Mary James

School context

Poppy County High School was a large, nine form entry, 11 to 18 mixed comprehensive of 1420 pupils and 82 staff in 1985, although the school roll had fallen to about 1280 in 1988 and staff numbers had likewise been reduced. The school served the rural town in which it was situated and a very large surrounding area from which pupils were bussed in. The locality was characterised by economic stability or expansion and the school roll was expected to rise again in the 1990s. Pupils came from mainly private housing and very few were from ethnic minority groups. As the only maintained secondary comprehensive in the area the school strongly identified with the local community which was a major point of reference. In contrast County Hall seemed far away, and indeed it was. This simple fact of geography had considerable significance for the RoA project in the school and helps to explain both the somewhat idiosyncratic initial development and tensions with the county project team.

Overview of the school RoA project

Interest had been taken in developing a RoA system a year or so before the opportunity to join the Essex pilot scheme was offered. The system being developed by the WJEC had seemed particularly attractive so work was begun on developing a similar comment bank system. The county team was unable, at that time, to deflect the school from this course and in 1985/86 the system was introduced to all first and fourth year pupils. Two further year cohorts were added in the two following years and by 1988, recording was being carried out with all of year one through to year six.

The initial system required all departments to submit a selection of comments for a whole curriculum comment bank for each year group, reflecting the subject-specific, cross-curricular and personal achievements expected in their area. Selections from banks were discussed and signed with pupils in class time. A further selection of these selections of comments was chosen by pupils in association with form teachers for computer printout as interim summary reports to parents or final summative statements. The response to the first issue of summary statements of this kind, in July 1987, was almost entirely negative so the whole system was reviewed and revised by the new co-ordinator appointed at this time. From 1987/88, the comment bank approach was abandoned (although it was retained for some departmental records) and a Record Book system was introduced. The new pupil Record Books contained academic summary sheets focused on target-setting, a record of general competencies and a personal record — all written in free prose with evidence. This document was to serve as a formative record, an interim report to parents and the data-base for the summative document. As previously, emphasis was placed on teacher-pupil discussion of all records.

PRAISE case study evaluation

Data were collected in eleven school visits from December 1985 to November 1987. The school was conducting its own evaluation with a 5 per cent sample of first and fourth year pupils in 1985 with follow-up in subsequent years. In the light of this it was agreed that both the school and PRAISE should use the same sample and share data where appropriate. I chose in particular to look at the experience of the fourth year sample as they progressed into their fifth year. Six pupils in particular were chosen for close study and these were shadowed over two days in an RoA week when there was an opportunity to observe teacher-pupil discussion and recording and interview both pupils and staff observed. In addition relevant school documents were collected, meetings for parents and employers were attended and participants interviewed, questionnaires were sent to 27 second and fifth year parents; other staff, notably the two school co-ordinators, were interviewed.

Much of the account of development in Poppy County High School revolves around a small number of major issues and themes.

First, was the dominant concern in the early period to develop an administratively viable and cost-effective RoA system for whole school implementation. It was assumed that a computerised comment bank would fulfil these criteria but experience proved otherwise and a negative reaction to summary documents stimulated a total review and revision of the system.

A second set of issues concerned the management of change and the influence of personal style, institutional norms and organisational structures. In this case the early RoA system was closely associated with the commitment of the core project team to a particular approach and, despite evident problems, radical reappraisal had to await staff changes.

The experience at Poppy County High also posed a question over the ability of schools to adopt an experimental and development approach to pilot work and to engage in inter-institutional collaboration: the incentive to move quickly to the institutionalisation of a system that seemed appropriate for the school appeared difficult to resist. This created tensions with the county team who regarded the school as undermining its centre-periphery-centre model of development for the county as a whole. However, the school's radical turn-round towards the end of the pilot phase suggested that resistances, were in this case more personal than structural and therefore more open to influence.

Finally, the development at Poppy School illustrates some of the problems and possibilities of attempting to develop a whole school, whole curriculum approach to recording in a large institution. In this respect the introduction of one-to-one teacher-pupil discussion in relation to all records was undoubtedly the school's greatest achievement.

Main issues carried forward for across-site analysis

HYDRANGEA SCHOOL, ILEA.
Case study worker: Barry Stierer

School context

Mixed-sex comprehensive secondary school catering for 11 to 18 year olds. Approximately 1175 pupils on the roll; staff of 86.

Catchment area was one of mixed housing and economic decline. Both parents of over 17 per cent of pupils were unemployed. About 8.5 per cent of pupils came from ethnic minority families.

Overview of the school RoA project

One of 20 pilot schools in the ILEA RoA pilot scheme.

The main components of the school's scheme were:

(i) The emphasis of the project was upon departmental development of Units and Unit Credits, with support and guidance from the school co-ordinator and from the Development Officer from the authority's central RoA team assigned to the school for one day per week throughout the pilot period. In general terms, departments were encouraged to develop Units and Unit Credits along the lines set out in the report by the Hargreaves Committee, *Improving Secondary Schools*, which advocated the re-structuring of the curriculum into clearly-defined units of work and the production of Unit Credits as the medium for assessing, recording and reporting achievement in the context of these Units. The departments which took part in the project developed Units and Unit Credits largely in a manner and at a pace of their own choosing. Two faculties, Mathematics and Science, took part in projects developing graded assessment schemes.

(ii) Some form tutors in years 1 to 4 introduced methods of tutorial-based reviewing and self-assessment. Profile sheets were also issued termly to all pupils in years 1 to 3, on which pupils' progress was recorded by subject teachers on a 1 to 5 scale under 'skills headings' nominated by each department.

(iii) A summative RoA (The London Record of Achievement) was issued to fifth year pupils in the summer of 1987, containing a Student Statement written by the pupil at the end of year 5, a School Statement written by the pupil's form tutor, summative departmental Credits, samples of 'best' work and other certificates of achievement.

PRAISE case study evaluation

Fieldwork for this case study was more limited than that for most other PRAISE case study institutions. For a number of reasons mainly centring on the teachers' dispute, fieldwork did not formally start at Hydrangea School until the autumn of 1986, nearly twelve months after fieldwork had begun for other case studies. This dispute continued in London well after near-normal working had been resumed in many other parts of the country. As a result, some visits in 1986/87 were cancelled at short notice due to last-minute decisions to close the school, making it difficult to sustain a continuity in fieldwork visits. Added to this was the fact that, from the middle of the spring term 1987, particular attention was devoted to the school's efforts to issue a summative RoA (The London Record of Achievement) to fifth year pupils in the summer, mainly because of the emphasis being placed on this element by the school, which made it difficult to document other developments within the school. Six visits were made to the school for the purposes of data-collection between September 1986 and May 1987. No new evidence has been collected since the summer of 1987.

Nevertheless, an attempt was made to document the range of development work undertaken by staff, by interviewing key members of staff, mainly heads of departments, heads of faculty and heads of year. Observations were carried out of several end-of-unit, one-to-one review sessions between the Head of Modern Languages and fourth year students of German.

186

With respect to the summative London Record of Achievement, observations were made of students discussing and preparing their Student Statements during tutorial lessons. One fifth year tutor was interviewed, as were six fifth year pupils after their portfolios were complete.

The short fieldwork period, the lack of fieldwork continuity, and the shift of emphasis within the study, resulted in an especially partial and fragmentary database. As a result, it has not been possible to produce a comprehensive case study report, even for the year 1986–87. Instead, this account relates almost exclusively to (a) the development and use of Units and Unit Credits in a small number of departments or faculties where the evidence collected was reasonably coherent and complete, and (b) the preparation of the London Record of Achievement for fifth year pupils in the summer of 1987, including the development and use of summative subject Credits.

The development and implementation of Units and Unit Credits in subject departments and faculties. This is a theme which runs throughout the report, and is addressed from several different perspectives.

Main issues carried forward for across-site analysis

The production of the summative London Record of Achievement in the summer of 1987, and in particular the Student Statement, the School Statement and summative subject Credits issued by departments.

The correspondences and differences between the terms 'cross-curricular skills', 'personal qualities' and 'Aspects III and IV' in the way in which they were used and in the kinds of characteristics to which they were understood to refer.

The issues comprising the debate over whether to develop a 'Cross-curricular profile of work-related abilities'.

The model of pupil self-assessment adopted for Unit Credits.

The conflict between the information needs of form tutors and the reporting needs of departments vis-a-vis Unit Credits.

Issues arising over the preparation by fifth year tutors of summative School Statements for the London Record of Achievement, and in particular the constraints imposed by the principle of positive-only reporting.

The diagnostic and target-setting function of end-of-unit reviews between pupils and subject teachers, as exemplified by six observed interviews.

The possible conflict between the function of formative Unit Credits as a means of communication with parents and the principle of pupil ownership.

The possible conflict between the inclusion of formative Unit Credits in the London Record of Achievement portfolio and the principle of positive-only reporting at the summative stage.

Issues arising from the development and implementation of summative subject Credits, and in particular the resistance of some departments to the notion of Summative Credits, and the difficulty some staff encountered when attempting to write positive comments about some pupils.

BUTTERCUP R.C. HIGH SCHOOL, LANCASHIRE
Case study worker: Susan McMeeking

School context

Buttercup R.C. High School was an 11 to 16 mixed comprehensive situated in an affluent Lancashire seaside town. In September 1985 there were 590 pupils, drawn mostly from private housing, 35 full-time members of staff and one part-time teacher.

When the school went comprehensive, it had to compete publicly and favourably with other schools in the area, especially the local grammar school which had a high intake of pupils. The diet at the school was described as being fairly academic and the school had a considerable reputation for getting good academic results but was also recognised as a school catering for special educational needs pupils. The school population was described as being positively skewed with a few low attaining pupils, and so did better in terms of academic attainment than other schools. The ethnic minority population at the school was low at less than 5 per cent.

The local area was one of economic stability with lower than average unemployment. There was more than one large employer in the area.

Of the 131 fifth year pupils leaving Buttercup at the end of the 1986/87 year, 60 pupils went on to futher education, 31 found employment and 29 went onto YTS courses. Eight pupils were registered as unemployed and the destination of three pupils were unknown.

Overview of the school RoA project

Buttercup was required to trial the years four and five version of the City and Guilds of London Institute School profile. Features of the scheme were pupil logbooks consisting of white and green pages. White pages were intended to be used as a place where pupils noted down their best achievements both in and out of school and then break these down into cross-curricular skills, the same skills that appeared on a 'progress profile'. This progress profile was completed in reviews between pupils and reviewing tutors which as well as recording achievements in relation to a list of cross-curricular skills (the checklist), also had space for the pupil's main activities and experiences to be recorded as well as for a target to be set. The green pages of the logbook were where the pupils recorded their private reflections.

The City and Guilds scheme operating at the top end of the school, was changed little during the pilot project at Buttercup, however, the school did begin to develop its 'own' scheme starting with first year pupils, drawing on the best elements of the City and Guilds scheme and also including subject input. This scheme began operating with first year pupils in September 1987 and it was intended that it would eventually replace the City and Guilds scheme at the top end of the school.

PRAISE case study evaluation

(i) In relation of the City and Guilds scheme

Fieldwork began in Buttercup in 1985. Because pupils of all abilities were not distributed evenly amongst reviewing tutors, two fourth year tutors were selected for close focus study during 1985/86 and 1986/87. In September 1986 a third tutor was identified for close focus study and observed during the rest of the period of evaluation. A number of review sessions between the three tutors and pupils selected for close focus were observed during the course of the case study and follow-up interviews carried out with staff and pupils. It was also possible to carry out interviews with other pupils involved in the scheme, other reviewing tutors and non-involved staff.

Following the issue of the summative documents for fifth year pupils in 1987, close focus pupils were interviewed and where possible, questionnaires sent to employers and further education institutions. Questionnaires were also sent to the parents of all close focus pupils. During the course of the evaluation interviews were carried out with the head, the other members of the senior management team, and with the school co-ordinator.

School and county-based INSET sessions, meetings of the Lancashire school co-ordinators, a meeting of reviewing tutors, a parents' evening for the initial cohort of pupils, and the summative document presentation ceremony were also attended by the evaluator.

(ii) In relation to school's 'own' scheme

The late emergence of Buttercup's scheme meant that the evaluation of development was less detailed than the evaluation of the City and Guilds scheme and was carried out through documentary analysis and interviews with the school co-ordinator.

The school's concern to develop a scheme that was better suited to the needs of it's pupils than the City and Guilds scheme in the sense that the emphasis was on the formative process rather than on the summative document. This was an important issue in that it gave rise to the development of a school scheme beginning with first year pupils, initially implemented during 1987/88. This scheme co-existed with the City and Guilds scheme operating with fourth and fifth year pupils.

Main issues carried forward for across-site analysis

The reaction from pupils that the City and Guilds progress profile did not record 'worthy' accomplishments but rather run-of-the-mill activities.

The general lack of use of logbooks by pupils for recording their achievements, activities and experiences, and the lack of reference to logbooks by tutors in reviews.

SUNFLOWER HIGH SCHOOL, LANCASHIRE
Case study worker: Susan McMeeking

School context

Sunflower High School was an 11 to 16 mixed comprehensive school situated close to the centre of a large East Lancashire city. The school opened in 1966 as a secondary modern and went fully comprehensive in 1972. Prior to reorganisation in 1984, the school covered the 11 to 18 age range.

Sunflower was in a poor state of repair needing £750,000 spending on refurbishment and a decision was taken by the County Council to close the school site in August 1988 and amalgamate Sunflower with another local school. This decision rendered untenable long term aims of the RoA project and developments must be seen in this light.

In September 1986 there were 630 pupils on the school roll and 45 staff but the school roll was falling, dropping from seven form entry in 1985 to just four form entry in 1986. This situation was exacerbated by the decision to close the school site. Pupils were drawn largely from council housing including an area designated as one of the most deprived in the country. Many pupils were drawn from one-parent families and unemployment was high at 15 per cent. Forty per cent of the school population consisted of English born pupils from various ethnic minority groups mainly from the Indian sub-continent — India, Bangladesh and Pakistan, but some from Tanzania, Malawi and Uganda. There was a degree of racial tension in the school but this varied in intensity.

Sunflower had a high percentage of educational special needs pupils. In September 1985, 48 per cent of the intake were special needs pupils, a figure that increased to 70 per cent for the 1986 cohort of pupils.

Although the area is one of high unemployment and economic decline, 40 per cent of the 1985 school leavers found employment (some via YTS courses). Pupil aspirations were described as modest and the jobs pupils found were those requiring few qualifications. The engineering and shoe industries offered a lot of opportunities, but there were few offers from the traditional textile industry. Many pupils found work in the retail trade and others took up some form of community work. Approximately 30 per cent took up courses with the local tertiary college, 20 per cent went onto YTS courses and 10 per cent remained unemployed.

Overview of the school RoA project

Sunflower was required to trial the City and Guilds of London Institute's School profile adapted for years 1 to 3. This scheme was substantially changed during the course of the pilot project. The original scheme consisted of pupil logbooks for continuous recording, and a 'progress profile' completed in reviews between form tutors and pupils. The latter comprised a list of cross-curricular skills, e.g. talking and listening, writing and use of equipment, and space where an example of achievement in each area could be entered. There was also space for the recording of pupils' main activities and for the noting of where progress had been made, what the pupils had done well in, and what they should pay special attention to before the next review. Logbooks consisted of white and green pages. In the white section, pupils recorded information relevant to the information on the progress profile and the green pages were for personal reflections.

The progress profile was rejected after the first reviews with pupils and was replaced with review sheets designed to explore particular themes, e.g. year 1 review sheets were aimed at getting pupils to identify skills they were acquiring, although individual review sheets used during year 1 had different sets of questions. City and Guilds-style logbooks were rejected after a year and continuous recording in the pastoral area ceased.

End of year statements were produced by pupils throughout the pilot project, one of the aims of the reviews being to provide pupils with the experience of talking and writing about themselves as preparation for the production of this statement.

At the end of 1986/87 academic year, these end of year statements were issued for first and second year pupils as part of an interim RoA which also contained subject-specific RoAs developed during the project and a pastoral report.

Fieldwork began in October 1985. A first year form group was selected for close focus study during 1985/86, 1986/87 and 1987/88. A second first year form group was identified at the beginning of the 1986/87 academic year and followed during the rest of the evaluation project. Observations of review sessions between both form tutors and their pupils were observed and follow-up interviews carried out with tutors and the pupils concerned.

PRAISE case study evaluation

Other form tutors involved in the RoA scheme were also interviewed. Academic staff were interviewed on two separate occasions during the course of the evaluation. Heads of faculty were initially interviewed in July 1986 when the development of subject-specific assessment schemes was still in its infancy. A more extensive programme of interviewing was carried out in July 1987 following the issue of the first interim summative statements.

During the course of the evaluation, interviews were carried out with the headmaster, other members of the senior management team and the school co-ordinator.

Other events attended by the national evaluator were planning sessions involving the initial group of first year form tutors, county-based INSET sessions and meetings of the Lancashire school co-ordinators.

Questionnaires were sent to the parents of all close focus pupils during the 1987/88 academic year.

An important issue at Sunflower was senior management's reasons for seeking involvement in the pilot project which was taken with a view to conducting a whole school review, and that RoAs would be seen as an integral part of a whole school policy to develop the curriculum. Whilst the head and the school co-ordinator felt that this was entirely successful (in June 1988), staff interviewed during the project expressed reservations about the introduction of the initiative. It must be said that it was not possible to explore any changes in view at the end of the period of the evaluation, so the success of the initiative in the eyes of other members of staff cannot be assessed.

Main issues carried forward for across-site analysis

The method of dissemination of information about the initiative at Sunflower during the project's history.

The abandonment of continous recording in the pastoral curriculum.

CAMPION UPPER SCHOOL, SUFFOLK
Case study worker: Mary James

School context   Campion School was a 13 to 18 mixed comprehensive upper school of 875 pupils and 54 full-time staff in 1986. The school was experiencing rapidly falling rolls and the demographic trend had been accelerated by the loss of a feeder middle school in a redrawing of catchment boundaries. During the period of study Campion drew its pupils from two middle schools, both serving the Suffolk town in which it was located and surrounding villages. The falling roll had influenced staffing and, although the Senior Management Team changed around the time of this study, many other staff were of long standing.

The medium-sized rural town in which the school was situated drew pupils from both private housing and council housing estates, but mainly the former. This should not however be taken as an indication that the children were mainly middle-class; indeed, many of the most local children came from low-paid families. Very few however came from ethnic minority groups. The town was dominated by one large local industry characterised by many small employers engaged in interdependent businesses. In addition, employment opportunities were provided in retailing, in the usual service industries and in a few small electronics and engineering firms. Despite the loss of a large manufacturer, the town was experiencing economic growth and unemployment was much lower than the national average.

Overview of the school   Campion Upper School had been involved with profiling for some years prior to
RoA project   both the advent of the Suffolk Record of Pupil Achievement (RPA) and the ESG funding of pilot schemes. School generated summative statements had first been issued in 1983, followed by Suffolk RPA Statements in 1984, and at the start of the DES pilot all pupils in years 3 to 5 were already involved.

Throughout the period of study, the Suffolk Record of Pupil Achievement (RPA) scheme was principally concerned with personal development and consisted of personal recording (in the 'profile') by pupils in the areas of personal qualities, out-of-school interests and activities, and general experiences of school and work. (At Campion School this activity was located in its established Guidance Programme which was taught by form tutors and occupied a total of three 50-minute periods per week.) Profiles were expected to be discussed periodically with pastoral tutors prior to final writing of 'statements' in the fifth year when a summary of pupils' profile records would be discussed and agreed, and tutors would incorporate their own Tutor's Review.

In line with the DES policy statement that recording processes should 'cover a pupil's progress and activities across the whole educational programme of the school' (paragraph 16) and that final summative statements would need to include both information of personal achievements and 'evidence of attainment in academic subjects and practical skills' (paragraph 17), Suffolk's original submission for ESG funds stated that the Suffolk Record was conceived as having three parts. In addition to the profile and statement referred to above, it was stated that 'skills profiles' in subject areas were being developed by the county's advisory service in consultation with teachers, and that these would be included in the record in the future. My evidence suggests that the intention to develop Skills Profiles at county level for incorporation in the existing Suffolk record was never fulfilled. Moreover, staff at Campion Upper School were not aware of any expectation that, in the context of this project, they should implement anything more than the profile and statement elements which had been in use since 1984. They never saw academic profiling as part of the Suffolk scheme, so they regarded the processes related to the profile and statement as the totality of the record of pupil achievement.

For this reason the initiative was almost exclusively associated with the pastoral curriculum and personal recording as it had been prior to the DES pilot phase. Indeed practice in the pilot phase was more or less continuous with what had gone before although the school benefited from additional resources to support statement-writing etc. During the pilot period the school, on its own initiative, began to explore the possibility of academic profiling and curricular provision for the development of cross-curricular skills. The school co-ordinator also became aware that the LEA team was 'looking at' similar areas. However, since the school had not been instructed by the county to develop or incorporate these elements into a more comprehensive record of achievement, it continued to regard its involvement with the pilot scheme as confined to the implementation of the existing RPA. In other words, staff were not aware of any expectation that the school should, at this point, **develop or pilot** a wider conception of a record of achievement along the lines of the DES statement of policy.

Furthermore, the school had suffered badly during the period of teachers' action, at which time it was also trying to implement a new curriculum structure, followed by GCSE and the new conditions of service. From 1985 to 1987, the period of my study, teacher morale was low and the school was suffering from 'innovation fatigue'; for these reasons senior staff took a deliberate decision to introduce no new changes for some years until the school was ready to move forward again. All these factors effectively limited the amount of experimentation and development that could be engaged with the consequence that the initiative at Campion School had the character of an implementation project rather than a pilot project.

Data were collected in ten school visits from November 1985 to November 1987. The main focus was the experience of pupils in a third year tutor group in 1985/86 who were followed into their fifth year in 1987/88. The same tutor remained with them throughout this period. This tutor group was observed in RPA-related Guidance periods and review sessions, and interviews were conducted with a sub-sample of six boys and girls and the tutor. In addition, interviews were conducted with identified 'key informants' amongst the staff, and relevant documents were collected.

PRAISE case study evaluation

As in all other institutional case studies conducted by PRAISE, the framework for analysis was developed from criteria in the DES policy statement as well as our own analysis of themes and issues which were emerging from our fieldwork in 22 institutions. In the case of Campion Upper School the effect of this was to create a report in which considerable space is given to discussion of what did not happen, especially with respect to academic recording. In other words, the initiative in the school was not simply evaluated in its own terms and against its own criteria but also in relation to the dimensions of experience in other settings. However, the absence of evidence in certain areas should not automatically be taken as a negative judgement; in a number of respects the reasons why something did not happen were well founded and worthy of further deliberation.

A number of composite issues emerged particularly prominently, although it should be pointed out that these were perceived to be issues by the evaluator. The school did not always share her perceptions.

Main issues carried forward for across-site analysis

First, as indicated above, the almost exclusive association of RPA with the pastoral curriculum and personal recording raised an issue concerning comprehensiveness. Although the school never considered that the initiative should cover or permeate the whole educational programme of the school, a lack of continuity with other curriculum, assessment and recording initiatives both pre-16 and post-16 was recognised and possible links were explored. However, no new development intended to co-ordinate various systems was implemented at school level during the DES pilot phase. On the other hand, concentration on personal qualities and achievements, as a deliberate counterbalance to the emphasis on academic achievement in most areas of school life, may in fact have been the greatest strength of the RPA in Campion School. Indeed it was recognised that the laudible aim to target and improve the poor self-image of many pupils, especially low attainers, could be undermined if, at national level, records of achievement were to be defined more widely to incorporate a large element of recording on the formal academic curriculum. In other words, it was thought that the personal element with the aim to promote personal development could become submerged once again.

A second issue arose in connection with the impression that RPA was regarded as a 'package' developed at county level for 'adoption' at school level. This perception inhibited creative adaptation but, more importantly, influenced the enthusiasm of teachers for the project. Some were grateful for the materials and guidance 'received' from the county, particularly at a time when they were expected to respond to numerous new initiatives. On the other hand, there were others who had little commitment to something in which they had no development role and which they regarded as an external imposition. Tutors' attitudes were unwittingly communicated to pupils whose response to RPA generally reflected the level of commitment demonstrated by their teachers. If pupils perceived that the RPA was considered important by tutors they valued it also, and vice versa.

In a different area, a third theme concerned the nature of tutor-pupil discussion. The opportunity to talk on a one-to-one basis was valued by both tutors and pupils and was another important contribution of the RPA. Moreover, observations of review sessions in Campion School suggested that discussions can be particularly fruitful when the tutor takes a low-key role which minimises his or her authority status. However, a remaining question concerned the relevance and validity of any attempt by form tutors to incorporate a necessarily brief and superficial review of aspects of academic achievement into a record that was chiefly personal.

HAWTHORN R. C. UPPER SCHOOL, SUFFOLK
Case study worker: Mary James

School context

Hawthorn Roman Catholic Upper School was a small 13 to 18 mixed comprehensive upper school of 510 pupils and 33 staff in 1985, falling to 450 pupils and 30 or 31 staff in 1988. In 1988, 86 pupils were in the sixth form. Both the church foundation and the small size of this school emerged as important in this study because both features promoted an atmosphere characterised by relaxed and friendly relationships between pupils and teachers and an emphasis on personal values.

The school was located in a town but drew many pupils from a very large, predominantly rural, catchment area. The town and its surrounding villages were experiencing a period of economic growth and there was a diversity of large and small local employers. Most pupils came from private housing but pupils from ethnic minority groups were very few. Staff perceived many parents to be ambitious for their children and the school was thought to have considerably more high attainers than low attainers.

Overview of the school RoA project

The school had been involved with the Suffolk RPA scheme from 1983 when it was introduced to fourth and fifth year pupils and incorporated into the school's Active Tutorial Work programme. However, in 1985 and 1986 implementation was badly affected by teachers' union action and the project had to be relaunched in autumn 1986. For this reason, the target group for the purposes of this study was the fifth year in 1986/87, although in 1987/88 all years became involved, including the third (intake) year.

The essence of the Suffolk Record of Pupil Achievement (RPA) scheme was personal recording (profiling) by pupils in the areas of personal qualities, out-of-school interests and activities, and general experiences of school and work. These personal records were expected to be discussed periodically with pastoral tutors prior to final statement writing in the fifth year when a summary of pupils' profile records would be discussed and agreed, and tutors would add their own Tutor's Review. Although the possible addition of a cross-curricular skills profile was discussed at both school and county level, nothing of this nature was developed or implemented at Hawthorn School during the period of study. Therefore the RPA was mainly associated with the pastoral curriculum.

PRAISE case study evaluation

Data were collected in thirteen school visits from November 1985 to February 1988. The main focus was the experience of one fifth year tutor group in 1986/87 although, for comparative purposes, there was some observation of another tutor group. The main study group were observed in RPA-related tutorial periods and review sessions, and interviews were conducted with a sub-sample of six pupils and the form tutor. Questionnaires were sent to 22 parents and a very small number of known users of summative statements. In addition, interviews were conducted with fifth year tutors, the school co-ordinator and the Head of Sixth Form. Relevant documents were collected and in-service sessions were observed.

Main issues carried forward for across-site analysis

Five themes or issues of particular interest and relevance emerged.

First, progress of the pilot at Hawthorn School was seriously handicapped by teachers' union action and difficulties arising from the unfortunate absences through illness of key members of staff. The account which follows should be read with this in mind. (Someone suggested that the illness was a research effect!)

Secondly, although purposes, materials and process principles were clearly defined and common to all participants, the implementation of recording processes varied considerably among tutor groups. This was recognised by staff and was a reason for their suggestion that I should observe review sessions in, at least, two tutor groups. Factors influencing diversity seemed most closely connected with the amount of preparation and the degree of time, energy and priority the tutor was prepared to give to RPA activities, which in turn related to the nature of other commitments, tutors' professional identities, and the character of relationships with pupils.

Thirdly, the school had a large proportion of high attaining pupils and staff suspected that these would ascribe little value to RPA. Whilst pupils stated a preference to use the time on academic work they nevertheless recognised the value of reflecting on their all-round achievement as persons and some thought the exercise had indeed contributed to their personal development. Their willingness to accept the emphasis of RPA seemed to be closely related to their acceptance of the culture of this church school which stressed personal, social and Christian moral values.

Fourthly, RPA was located in Active Tutorial Work in pastoral periods which provided recording processes with a supportive framework. However, some anxiety was expressed that RPA, with its outcome in a summary statement, introduced a product orientation that might eventually undermine the ATW principle that certain processes have intrinsic value.

Finally, a major concern was the lack of obvious continuity between RPA and other assessment and recording initiatives both pre-16 and post-16. Both within the school and at county level the need for co-ordination was recognised but no new development was implemented at school level during the DES pilot phase. There was also a worry that the valued emphasis on personal achievements and qualities might be lost if the RPA were to become a comprehensive RoA covering all aspects of academic achievement.

JASMINE HIGH SCHOOL, WIGAN
Case study worker: Susan McMeeking

School context

Jasmine was an 11 to 16 mixed comprehensive school situated in a suburb of Wigan. It was a comparatively new community school fought for by the lcoal residents. The Wigan definition of a community school was the more efficient use of buildings and the school offered a number of courses for the local community and had as many pupils in the evening as it did during the day.

In September 1985, there were 695 pupils and 45 members of staff. The school had an ethnic minority population of less than five per cent. The school population was largely drawn from private houses.

The area was one of economic decline and above-average unemployment with no large employers in the local area.

Overview of the school RoA project

The RoA scheme at Jasmine developed from scratch with the emphasis being on the development of RoA schemes within individual departments. During the latter half of the evaluation, there was a strong move within a number of departments towards the unit accreditation scheme operated by the Northern Partnership for Records of Achievement (NPRA), Wigan being a participating authority in this very extensive scheme.

It was originally envisaged that departments would develop, assess and record personal qualities as well as subject-specific skills. It was later felt that this was asking staff to take too much on board (although it was recognised that personal qualities could be included in subject-specific RoA schemes). With this change in view came the implementation of a pastoral scheme operating in registration time for first and second year pupils. In diaries pupils recorded details of their in-school and out-of-school activities under various headed sheets. The aim of the diaries was to provide pupils with the experience of recording information, but perhaps more importantly, encouraged them to examine their personal achievements.

PRAISE case study evaluation

During the course of the evaluation, interviews were carried out with the headmaster (on several occasions), the school co-ordinator (on several occasions) and with heads of department (on two occasions).

Samples of second and third year pupils were identified and review sessions observed in connection with a PE/expressive arts course on 'The Seven Ages of Man' and in home economics. Follow-up interviews were carried out with both staff and pupils.

A first year registration period was observed during which pupils filled in their diaries.

Several types of meetings were attended, namely, departmental and whole-school INSET sessions, a meeting of heads of department to discuss RoA developments, and a weekend meeting of school governors held to discuss significant curriculum initiatives including RoAs.

A postal questionnaire was sent out to parents of the pupil sample.

At LEA level, meetings of the Wigan school co-ordinators were attended, and an employers' conference.

Main issues carried forward for across-site analysis

The adoption of NPRA's unit accreditation scheme by a number of departments.

A major impact of the RoA scheme at Jasmine was the creation of the posts of curriculum co-ordinator which replaced heads of year whose job it was to examine the recording of achievement in each year and to identify what pupils in each year should experience. This post was regarded as one that married together academic and pastoral concerns. The school co-ordinator became Curriculum Co-ordinator for Jasmine High School which was reported to reflect the school's commitment to a whole school policy of curriculum development.

Another major impact was that the recording of achievement became the umbrella under which curricular developments were co-ordinated and became the key to the school's 'map of territory'.

MARIGOLD HIGH SCHOOL, WIGAN
Case study worker: Susan McMeeking

School context

Marigold High School, Wigan was a mixed 11 to 16 comprehensive school situated in a suburb of Wigan. The school was originally a secondary modern school and had retained some teaching staff from this period. The school grew steadily in size following reorganisation in 1971 and at the time of writing there were 1083 pupils on the school roll and 65.5 staff equivalents. The ethnic minority population of the school was low at less than five per cent.

The school was situated in an area of economic decline with unemployment above the national average. Approximately 10 per cent of the school leavers found work, 40 per cent went on to futher education and 50 per cent to YTS courses.

Overview of the school RoA project

In registration periods at Marigold, pupils made entries on a series of headed sheets ('attendance', 'uniform', 'merits', 'service to others', 'reading and writing', 'visits and expeditions', 'speakers', 'TV/Radio programmes', 'hobbies', and 'teams and clubs') which were designed to throw light on their personal qualities. Pupils' in-school and out-of-school activities and achievements were recorded on these sheets.

Subject specific RoAs were developed and issued as part of an interim RoA which went home for first year pupils in July 1987 along with the 'About Me' summary sheet which was based on the continuous recording that took place in registration time.

PRAISE case study evaluation

Fieldwork began in November 1985. During the course of the evaluation interviews were carried out with the school co-ordinator, the headmaster, one of the deputy heads, heads of department, other academic staff and form tutors involved in the pastoral side of the initiative.

In September 1986, a sample of first year pupils was selected for close focus study. Interviews were carried out with these pupils on several occasions. During observations of registration periods, discussions with other pupils was possible.

Events attended by the national evaluator included an INSET session held for all 1986/87 first year form tutors in July 1986, several of the RoA steering committee's planning and discussion sessions, and a parents' evening held for the parents of first year form tutors in December 1985.

At LEA level, meetings of the Wigan school co-ordinators were attended and an employers' conference.

Main issues carried forward for across-site analysis

The change from direct to indirect recording of details of pupils' personal qualities.

The influential role of the RoA Steering Committee on the development of both the pastoral and the subject-specific areas of the initiative.

The desire of academic staff to issue a traditional style report to report on progress in GCSE and the potential this had for undermining the initiative.

The emphasis on in-house development and the perception of staff that consulting other RoA schemes was 'not allowed'.

The adoption of Marigold's system of continuous personal recording by one of its feeder primary schools.

BEGONIA UPPER SCHOOL, EMRAP
Case study worker: Susan McMeeking

School context

Begonia was a 13 to 18 mixed comprehensive school situated on the outskirts of a large city in the East Midlands.

The school opened in 1962 as a secondary modern and was formally opened as a comprehensive upper school in 1974 following an extensive rebuilding programme. In September 1986 there were 1170 pupils on the school roll, 130 of whom were in the sixth form. At this time there were 80 members of staff. The ethnic minority population in the school was low at less than five per cent.

The area around the school was one of economic stability but with above-average unemployment. The pupils were drawn largely from council housing. The school had an equal mix of boys and girls and a spread of the whole ability range and social mix.

Thirty per cent of fifth years stayed on into the school's own sixth form, between 15 and 20 per cent went onto the FE college, 35 per cent found employment, and the rest of the pupils took up YTS schemes. As far as employment was concerned, a lot of the pupils went into the service industries, shoe industries and light and heavy engineering industries.

Overview of the school RoA project

Begonia developed its RoA project from scratch. The scheme consisted of a pastoral 'attitudes and relationships' document completed in reviews between form tutors and pupils. Assessment of personal qualities made by both staff and pupils appeared on this document (with supporting evidence on the fourth and fifth year version of the document). Other information recorded included pupils' outside interests and hobbies and responsibilities, details of voluntary and paid work undertaken; both form tutor and pupil produced a statement giving a summary of progress and action needed.

Subject staff identified assessment criteria which, for the 1985/86 cohort of third year pupils, were listed on individual subject RoAs along with a teacher comment and a list of cross-curricular skills. The latter were removed after the first issue of these documents. The second version of the third year academic RoA included space for pupil comment as well as a teacher comment, although this was only completed by one third year form group who were selected for involvement in subject-specific reviews following the LEA co-ordinator's expression of concern that without teacher-pupil discussion, the academic RoA documents were not RoAs.

Computerisation of academic RoAs began in 1986/87 and by December 1987, all academic RoAs for years 3 to 5 were produced on the computer (although different systems were in operation). Computerisation was one factor that affected the development of academic RoAs during the course of the pilot project at Begonia and was instrumental in bringing about the change from grade-ringing and tick-box systems to comment banks.

The 1987/88 cohort of fifth year pupils were the first to leave with a summative document in March 1988. Begonia conformed to the standard EMRAP format.

PRAISE case study evaluation

During the pilot project at Begonia, two pupil samples (one from the 1985/86 cohort of third year pupils and one from the 1986/87 cohort) were identified and observed in both pastoral and academic reviews.

Heads of department were interviewed on two occasions in order to keep abreast of developments with regard to the subject-specific element of the RoA project.

The 1986/87 cohort of third year form tutors were interviewed after their first pastoral reviews with the pupils in their form groups.

The school co-ordinator was interviewed on several occasions and the headmaster was also interviewed.

A questionnaire was sent to the parents of the pupils in the sample.

A meeting at which the LEA co-ordinator for RoAs addressed heads of department was attended.

County-based INSET sessions were also attended.

The removal of the cross-curricular section from subject-specific RoA documents.

Main issues carried forward for across-site analysis

The computerisation of the RoA scheme.

The issues of finding time for the development and implementation of the RoA initiative.

GERANIUM SCHOOL, EMRAP
Case study worker: Susan McMeeking

School context

Situated in a suburb of a large city in the East Midlands, Geranium School was an 11 to 18 mixed comprehensive with 1730 pupils and 98 staff. The school consisted of an upper block where fourth, fifth and sixth form pupils were taught and had their form groups and a lower block where first, second and third year pupils were based. Although the school was on a split site both sites were located in the same grounds.

The percentage of ethnic minority pupils was small at less than five per cent, the pupils in this group being mainly from Punjabi and Chinese families with good English.

Pupils were drawn from a roughly equal mix of private and council housing. Some of the former was described as being of poor quality. A lot of pupils were drawn from a large modern council estate surrounding the school. Some were drawn from within half a mile of the city centre where property was very poor-quality terraced housing.

The school's catchment area was described as unfavourable for pupils staying on for further education. An analysis of the destinations of the 1985/86 cohort of fifth year leavers revealed that over 50 per cent of the pupils took up YTS courses, a greater percentage than for the authority as a whole. Less than 23 per cent went into further education or into the sixth form at Geranium (a joint venture with the local college of FE), although further training of high quality was provided by some of the engineering companies based in the local area. Twelve per cent of leavers found employment although because of the location of several large engineering companies in the city, boys seemed to have more success in finding permanent employment than girls.

The local area was one of above-average unemployment but with a stable economy and with several large employers.

Overview of the school RoA project

The recording system developed at Geranium had two important features, namely, diaries that were introduced into the Personal and Social Education (PSE) programme, and an interim RoA that was first issued instead of a traditional report for first year pupils towards the end of 1986/87 and which was intended to replace reports for other year groups as the project moved up through the school.

Diary work involved reflection on various personal skills (e.g. note-taking) and personal qualities (e.g. ability to work in a group) relating to the content of the PSE course and was also the place where pupils reflected upon their academic experiences. All departments in contact with the first year pupils were involved in devising subject-specific RoA schemes. These subject-specific schemes formed the bulk of the interim statement but on this document there was also space for: a free pupil comment, where the target form group (i.e. the group followed closely by the researcher during the period of investigation) were asked to reflect on their academic experiences as well as upon their social relationships; form tutor assessments of personal qualities, e.g. co-operation and attitude to others; and, lists of pupils school-based and community-based activities.

PRAISE case study evaluation

Fieldwork began in October 1985. During the course of the evaluation, interviews were carried out with the headmaster, the acting headmistress (during the headmaster's secondment), deputy heads, heads of department and other members of academic staff, first and second year form tutors involved in the teaching of PSE, and form tutors involved in the implementation of the PSE diaries. Observation of PSE lessons before and after the introduction of the diaries was also carried out.

In September 1986, a sample of pupils was selected for close focus study. These pupils were interviewed on several occasions during the evaluation. Pupils were not observed in reviews in either the pastoral or the academic curriculum as these were unplanned.

Questionnaires were sent to the parents of the close focus pupils and it was also possible to interview a number of parents at a parents' evening for the parents of 1986/87 cohort first year pupils.

Events attended by the national evaluator included a meeting of all heads of year when the RoA initiative was explained to them, a parents' evening held for the parents of the 1986/87 cohort first year pupils, and school-based, LEA-based and Consortium-based INSET sessions.

The establishment of the PSE diaries and their use by form tutors as a means of establishing a dialogue with pupils.

Main issues carried forward for across-site analysis

The lack of enthusiasm from some members of staff for the initiative.

The position of the school co-ordinator and the need for a deputy head to sign RoA related documents going out to staff in order to give the school co-ordinator sufficient 'clout'.

The persistence of norm-referencing in the interim RoA statement issued instead of a traditional report to parents.

The perception that the introduction of diaries into PSE had enhanced the status of this previously poorly regarded subject.

MAGNOLIA GIRLS' GRAMMAR SCHOOL, EMRAP
Case study worker: Mary James

School context Magnolia Girls' Grammar School, was an 11 to 18, four form entry, selective girls' school of approximately 716 pupils and 42 teaching staff. Selection from the top 30 per cent of the ability range was based on an 11-plus using Moray House tests plus a primary school assessment of suitability for grammar school education.

The town in which the school was situated had a population of approximately 32,000, living in private and council housing, some on extensive housing estates. One of these estates was rather run-down but there was nothing that could be described as slum-dwelling. Ethnic minority groups were small and amounted to a small Asian community and a few Chinese families working in the restaurant trade. The town was economically stable and unemployment was at, or below, the national average. There was little major industry in the town and most people found employment in the three medium-sized factories, in light manufacturing industry or in the service industries. The latter attracted many of those girls from Magnolia School who did not proceed to higher education or advanced training.

Overview of the school In joining the EMRAP consortium, the LEA had taken the decision to involve all
RoA project the schools and the FE College in this particular town. There were several reasons for this. The town had been somewhat neglected in terms of new initiatives and this was an opportunity to give them a boost. Also it was felt that requiring all schools in one area to be involved, whether they would have elected to or not, would constitute a genuine pilot for an initiative that was planned to become national. This choice of strategy explains why few schools in this pilot had profiling experience on which to base their development. And some, including Magnolia Girls' Grammar School, had at first little inclination to be involved. LEA and schools, and to some degree the consortium, started from scratch and the progress made during the pilot phase must be understood in this light.

In common with other pilot schools in the town, Magnolia was asked to begin pilot work with fourth and fifth year pupils. Therefore in September 1986 recording began with the fourth year cohort and this was continued in 1987/88 when the next fourth year group was also brought in. The first summative statements were therefore due to be issued in summer 1988 — after the end of PRAISE data collection.

Throughout the pilot phase the school never had much sympathy with the idea that whole new systems of teaching, learning, assessment and recording should be instituted as part of the RoA initiative. (Although such changes came about during this time they were more a direct result of the introduction of GCSE.) RoA development was therefore focused on improving the school's interim and summative reporting systems in line with the RoA principles that summary records should be positive, supported by evidence and discussed with pupils. To this end, existing procedures for reporting to parents were reviewed and revised. A fifth year summative document was also developed to include a record of courses and examinations taken, a record of academic achievement based on subject reports, a pupil record of personal achievement and experiences, and an overview statement agreed by both the pupil and her form tutor. A pupil statement and a parental contribution was also considered. This latter, together with efforts during the development phase to solicit the views of parents, reflected the importance attached by the school to parents as 'users' of such reports.

PRAISE case study Data were collected in nine school visits from November 1985 to November
evaluation 1987. One fourth year tutor group, and six girls in particular, were observed in

Life Skills lessons and in discussions with their form tutor in 1986/87. These observations were supplemented with examination of relevant documents, observation of one RoA-related staff meeting and interviews with key members of staff. The school co-ordinators were particularly helpful in providing detailed written records of the development over time.

Three clusters of issues of particular interest emerged during the course of this study.

Main issues carried forward for across-site analysis

First were the effects associated with the fact that the school had been directed by the LEA to become involved in this project, and the implications this may have for the introduction of a national RoA system. The tensions between the school team and the county team were particularly marked although other factors contributed to this which were themselves of interest with regard to the management of change.

Secondly, the initiative was defined within the school as being primarily concerned with the reporting system, rather than the whole of curriculum, teaching, assessment and recording. Whereas there was little evidence of curriculum change directly attributable to RoA, teachers' attitudes to the RoA initiative undoubtedly changed and the proposed summary document had the potential to fulfil many of the expectations outlined by those who, like the consortium team, defined the initiative more widely. The idea that RoAs should be an instrument for radical curriculum change was rejected because the school was reasonably happy that it was already satisfying the needs of its customers. The perceived relevance of RoAs in selective school contexts, except insofar as they improved reports to parents, was therefore in some doubt.

Finally, a group of issues emerged regarding pupil differences. These related specifically to a possible gender difference in that underestimation of achievement by girls was especially marked, and that high attaining girls were more interested in the judgement of their teachers than in having an opportunity to assess themselves.

ROSE SCHOOL, EMRAP
Case study worker: Mary James

School context

Rose School was an 11 to 18 mixed comprehensive school of 1056 pupils and 61 teaching staff (in 1985). In 1979 it was reorganised into a comprehensive from a grammar school of ancient charitable foundation which still provided it with a small additional source of finance and voluntary-aided status. Many of the grammar school staff continued to teach in the school and a predominantly traditional ethos persisted. At the time of the RoA pilot the school occupied a split site and was six form entry although rolls were falling and the school expected staff cuts. Few pupils came from ethnic minority groups and severe social problems were rare. The main problem identified by the Head — one often associated with rural areas — was that a proportion of pupils appeared poorly motivated, apathetic or lethargic.

The school was located in a small town (population 14,000) although many pupils came from nine or so surrounding rural villages. Most occupied private housing. The area was economically stable and most employment was provided by agriculture and its related industries and services although two energy industries were large employers and there was some light industry.

Overview of the school RoA project

The school's staff had expressed interest in profiling prior to the invitation to join the EMRAP scheme although nothing had been implemented at the time. Subsequent development was mainly school-focused though broadly based on agreed EMRAP principles and the DES statement of policy. A major decision was to begin work with first year pupils and form tutors, who had experience of Active Tutorial Work, and develop a RoA system which could then work through the school. More attention during the pilot phase was therefore given to recording processes and the modification of school reports to parents in line with EMRAP principles than to the details of summative documentation.

During 1985/86, all first years were involved in personal recording based on a structured diary. In 1986/87 personal recording continued with this cohort but the new first year group was introduced to subject assessment and recording as well. In 1987/88, the new first year cohort was again involved in both personal and academic recording and the previous two year groups continued as before, except that the pastoral component with the current third year was based less on diary work and more on recording associated with decision-making skills. At this point the school also made a successful application to the LEA for accreditation of a limited summative document for the current fifth year which would simply contain a 'personal statement' from pupils and an agreed 'statement of personal qualities and skills'. The school hoped to issue a summative document of this kind in the interregnum until 1991 by which time the first year target group in 1986/87 would reach the fifth year and expect a full summative statement of both personal and course achievements.

PRAISE case study evaluation

Data were collected during ten school visits from October 1985 to February 1988. A main focus was the experience of one first form tutor group in 1985/86, and another in 1986/87, and their subsequent RoA experience as they progressed up the school. Three boys and three girls from each group were chosen for special study, which involved observation of lessons and review sessions, interviews and scrutiny of records produced. In addition, interviews were conducted with key project staff and those with whom pupils came in contact during two days of shadowing. Questionnaires were also sent to one form group of parents. Numerous relevant documents were collected and in-service sessions were observed.

Main issues carried forward for across-site analysis

Three dominant themes each with a number of dimensions emerged prominently in this school.

First was the relationship between personal, cross-curricular and subject-specific recording and the implications for the dual pastoral and academic curriculum organization. A prominent management objective was to break down the pastoral/academic divide and there was evidence both that subject achievement was a focus for pastorally-based discussion and recording, and that departments were moving towards the assessment and recording of personal and social skills as an important element in subject achievement. However, this created problems most of which were associated with the degree to which staff were equipped to deal with areas outside their knowledge, experience or competence.

A second major theme concerned issues surrounding the development of a school-wide reporting policy based on RoA principles. From summer 1987, interim RoA summary statements became synonymous with reports to parents and these were expected to be positive, provide concrete evidence of achievement and be jointly agreed by pupils and teachers. However issues of ownership (parents, pupils or school) and disagreement over the exclusion of negative statements at this stage loomed large.

Finally, the management of the innovation raised a number of issues. Most had to do with the best way to promote development which those expected to implement change would be committed to, whilst ensuring that certain common purposes and principles are achieved. This is a recurrent theme in innovation research but in this context it emerged as a tension, within the school, in terms of uniformity and diversity in departmental patterns of development, and, between the school and the LEA, in terms of appreciation for freedom of school-focused development but also a felt need for clear leadership.

Other themes of particular interest related to pupil self-assessment, a formative-summative tension in one-to-one review sessions, implications of recording with younger pupils, and some concrete evidence of increases in pupil motivation.

COLUMBINE MIDDLE SCHOOL, OCEA
Case study worker: Barry Stierer

School context   Mixed-sex comprehensive middle school catering for 9 to 13 year olds. Approximately 750 pupils on role; staff of 42.

The school's catchment area covered roughly half of a small market town and included about an equal mix of private and council housing. Less than 5 per cent of pupils came from ethnic minority families. The local area was one of economic decline, although local unemployment was not above average.

Overview of the school RoA project   A pilot school within the consortium of local education authorities piloting the Oxford Certificate of Educational Achievement (OCEA).

Main areas of activity:

(i) The school piloted the OCEA P Component and the OCEA Science G Component.

(ii) Trials of OCEA English G Component materials were carried out by the English Department on an 'Associate School' basis.

(iii) Other areas of relevant work not directly subsumed by OCEA but which came to light during the evaluation include (a) the introduction of new pastoral programmes and structures in the junior division of the school (years 1 and 2) in 1986/87 and (b) the development of 'learning logs' in the Art Department through an affiliation with the SCDC National Writing Project.

(iv) Target groups: 1985/86 — All classes in year 3 (11-12 year olds)
                 One class in year 4 (12-13 year olds)
       1986/87 — All classes in years 3 and 4

PRAISE case study evaluation   Fourteen visits were made to the school over 15 days for the purposes of data collection. The main focus of study was the development of the OCEA P Component and the Science G Component, initially with 11 to 12 year olds in 1985/86. A single class of pupils in this first cohort was studied closely, and used as a vehicle for documenting both sides of the OCEA project. This involved observation of the meetings of the team of tutors developing the OCEA P Component; observation of Science departmental meetings; observation of tutorial lessons and Science lessons for the sample of pupils; observation of tutor-pupil interviews; interviews with pupils and with form tutors, year heads, Head of Science, other members of the Science department, and the school RoA co-ordinator. An attempt was also made to keep abreast of other related strands of activity through regular discussions with relevant members of staff.

Main issues carried forward for across-site analysis   The difficulties encountered by the Science Department in implementing the assessment framework of the OCEA Science G Component, despite its commitment to new forms of teaching and assessment.

The difficulties in sustaining continuous recording in 'Think Books', and the greater success of alternative forms of recording in which the format, purpose and focus changed frequently.

The way in which the implementation of the OCEA P Component was felt to be dependent upon the in-house development of a comprehensive tutorial programme.

CORNFLOWER SPECIAL SCHOOL, OCEA
Case study worker: Mary James

School context

Cornflower School was a non-residential special school for physically handicapped and delicate boys and girls aged 4 to 16-plus. Many had serious illnesses, multiple handicaps and additional learning difficulties. From September 1985 to March 1987 the school roll fell from 75 to 62, largely as a result of changes in statementing of children with special needs. In 1985/87, the school also had 9 teaching staff, 4 ancillaries and 2 nursery nurses. By 1988, no pupils leaving at 16 went into employment; most went on to mainstream or special further education.

The school was opened as an Open Air School in 1930 in a recreational park in the suburbs of a city which, as a chief centre for the textile and clothing industry, has experienced a period of economic stability. The city has a large ethnic community, mainly of Asian origin, which constitutes at least 30 per cent of the total population. Most pupils came from the city and its suburbs although some came from further afield since the school served the whole of the county.

Overview of the school RoA project

The school was accepted onto the OCEA pilot without having completed a preparatory programme. Approximately £4,500 of funding was provided annually for use in 1985/87 (the pilot phase defined by OCEA) but this was reduced to near normal level by March 1988 (the end of the DES pilot phase). The school chose to use these funds to supply basic equipment for a science laboratory and a computer room and to employ a full-time clerical assistant. Extra resources were found to provide supply cover in support of a Teacher Fellowship for two terms and to employ a further clerical assistant for a term.

The school began by piloting the OCEA P Component and the G Component in Science with all the nineteen 13 to 16 year olds who formed two 'classes' (a more familiar term than 'tutor group' within the school). In the second year work on Maths G was introduced. As in other schools which began OCEA pilot work in 1985, the terms 'P' and 'G' persisted although in schools which joined OCEA in 1986 P became better known as the 'reviewing and recording' component and G became known as 'curricular assessment frameworks or models'. P and G are retained in this account because these terms were used in the school for most of the pilot phase. In July 1987 four pupils from Cornflower School received summative statements. When in 1987/88 funds from OCEA were no longer available the emphasis changed and most effort was directed towards instituting a system of reports on all pupils of all ages in every lesson across all subjects. Whilst P summaries were still produced as a result of teacher-pupil discussion, personal recording by pupils was no longer a prominent feature.

PRAISE case study evaluation

Data were collected during ten school visits from December 1985 to November 1987. The main focus was the experiences in relation to P and G recording with the nineteen 13 to 16 year olds initially involved in the project, with special attention given to three girls and three boys. Lessons and one-to-one discussions were observed, pupils were interviewed, and relevant documents were examined. Formal interviews were also arranged with all teaching staff, bar one. OCEA staff meetings, an in-service session and a parents' evening were also attended. Questionnaires were sent to parents and users who had sight of summative statements in 1987. Interviews and informal discussions were also arranged with the LEA co-ordinator, the special needs adviser and a senior LEA officer.

Main issues carried forward for across-site analysis

Serious illness, and even death, was an ever present reality in Cornflower School and the notion of a record of achievement must be understood in this context. Indeed, coming to terms with progressive disability and mortality was perhaps the most significant achievement for many pupils although making a record of such achievement seemed inappropriate.

Other, more specific, issues associated with pupils' illnesses and handicaps also loomed large. For example, patterns in recording that appeared to be related to differences in age, gender and ethnicity were overlaid by other factors such as low attainment, low self-esteem, lack of maturity and general vulnerability which stemmed more directly from the special nature of multiple handicaps. In Cornflower School all these factors interacted to precipitate dilemmas concerning, for instance, the potentially negative effects of consistent non-achievement on assessment criteria designed for the whole ability range; ethical questions arising from the disclosure of social and emotional problems in the context of recording out-of-school experiences; the legitimacy of 'improving' pupil self-accounts in order to protect the individual and the school. On the other hand, Cornflower School was notable in providing some concrete evidence of improvement in pupils' motivation as a result of the OCEA initiative.

Another prominent theme was the influence of the Head and his overt use of the OCEA initiative as a vehicle in his plans to reorientate the curriculum and organisation of the school and to attract material resources. In any chicken and egg analysis, OCEA has to be seen as serving as much as initiating change.

The evidence relating to the use of ancillary staff is also interesting, and probably unique amongst our case studies. Of particular note was the role of the 'person other than a teacher' in shadowing pupils and recording her observations of pupils' achievements in the classroom whilst leaving the teacher free to teach. Although this was a very expensive resource, and there were some problems associated with a lack of professional expertise, the idea of collaboration between staff in the classroom in fulfilment of the multiple tasks of teaching, assessment and recording — and co-ordinating records — might usefully be developed further.

Finally, the experience of Cornflower School, highlighted a general issue relating to records utilising skills-based assessment frameworks. Whilst attempts to describe what pupils can do undoubtedly represents an important advance on single grades based solely on what they know, skills statements devoid of content and context are difficult to interpret particularly in terms of level of achievement. The indications are that some way has to be found to describe achievement **in context** succinctly.

PERIWINKLE SCHOOL, OCEA
Case study worker: Patricia Broadfoot

School context

Periwinkle was a mixed comprehensive and community school, situated in a rural area serving a number of local villages. The village itself had a population of 4,000 and the catchment area spanned an area of approximately 9 × 11 miles between a large university town and two rural market towns each with their own comprehensive schools. The school had suffered from falling rolls in recent years, having 1,000 pupils and 58 staff in 1985 falling to 760 pupils in 1988 — with a staff of approximately 48. There were some adults in the O and A level classes and one part-time FE tutor. The falling rolls had led to the possibility of smaller classes, approximately 25 in the short term, which had been a factor in promoting high morale in the school and which facilitated development of OCEA procedures.

The school was situated in an affluent area with many professional and university families as well as farming families. The building itself was bright and attractive, having been built in 1958 with a separate purpose-built block for first years in which classrooms were arranged round a large central resources area.

The school was divided vertically into Houses called after local figures for such purposes as games, swimming and athletics, but the evaluation did not pick up any relationship between this structure and OCEA.

Overview of school RoA project

The school piloted the Oxford Certificate of Educational Achievement within one of the four local education authorities involved.

Main areas of activity:

(i) The school piloted the OCEA P Component with first and fourth year classes in 1985/86. From 1986 onwards a tutorial programme involving the use of the OCEA P Component was in use for all year groups. All pupils were registered for OCEA accreditation of P and in 1986/87 all first and second years were registered for OCEA G in Maths, all the third year and some fifth years were registered for OCEA English, and all the first years for humanities. At the end of 1986/87 academic year, all pupils who had completed a summative statement were awarded a pilot OCEA Record of Achievement.

(ii) The introduction of OCEA in the school was marked by a major change in the reporting procedure to parents in which pupil self-assessment alongside teacher-assessment is now the defining feature.

PRAISE case study evaluation

Particular attention was paid to one class of pupils in the first year group who in 1985/86 formed the first pilot year group for the P Component. The field work in this respect involved interviews with a small sample of pupils in this class, observation during the weekly double tutorial period, observation of pupils recording and reviewing with their tutor in class, observation of the same pupils in humanities classes, which in this school provides for study in history, geography and English in the first and second years. Interviews were conducted with the tutor and with other members of staff involved with this tutor group. Attention has also been given to the fourth year pilot group who received the first summative OCEA leaving certificate in the summer of 1987 and observation of recording sessions in which pupils were preparing their final statement; interviews with year head and tutors for this group also constituted an important focus for the evaluation. Pupils were also interviewed before and after receiving their summative statement and completed a brief questionnaire. Interviews have been conducted with a number of the school's one-day secondees for science, modern languages, maths and English and for the P Component. Other interviews have been conducted with relevant heads of departments, heads of year, school co-ordinator and the headteacher. Particular attention has also been given to the piloting of the OCEA accreditation process in this school with a view to monitoring issues in accreditation and LEA development of the scheme.

Periwinkle School is one of very few that have experimented with school-based accreditation already. Whilst little can be said about the impact of such accreditation procedures on the quality of school processes as yet, the model proposed is an extremely interesting one which appears to have a lot of potential, incorporating as it does the lessons learned from school self evaluation in general. It also gets round the problem of punitive accreditation when it is not possible to dis-accredit an institution.

Main issues carried forward for across-site analysis

Periwinkle school demonstrates clearly the growing trend for records of achievement to be integrated into broader LEA developments. The recent commitment to appointing a co-ordinator to oversee these developments as a whole within the school reflects the need which is emerging in the PRAISE evaluation as a whole, for integration between records of achievement/GCSE and other current initiatives, such as TVEI.

The issue of preparation for innovation comes out clearly at Periwinkle school: although the tutorial programme appears to be developing much more satisfactorily now, it undoubtedly went through a very rough time when it was first introduced because teachers were not committed to it or did not understand the associated role of OCEA P. The lesson learned is that it needs to be a much more worked-through, grass roots development.

The fact that OCEA is still itself developing has created problems. Problems of continuity have been quite marked as each new batch of secondees starts work. Each year there is a noticeable sacrifice of cummulative experience as a result. This is perhaps one symptom of a more general issue within OCEA about the balance between school-based and scheme-based development work and to what extent externally-devised systems can be imposed effectively on a school.

Periwinkle school provides clear evidence of problems with the erstwhile G component of OCEA which are now being resolved in a variety of ways.

Impact of provision being made mainly for group recording and reviewing opportunities rather than on an individual basis.

Pupils' response to the reviewing process.

The development of assessment criteria within academic subjects and the impact of GCSE.

Responses to different INSET strategies.

WISTERIA SCHOOL, OCEA
Case study worker: Barry Stierer

School context Mixed-sex comprehensive secondary school in a suburban district of an industrial city, catering for 11 to 18 year olds. Approximately 1300 on roll; staff of 80.

Mixed catchment area, with pupils predominantly in private housing. Although unemployment in the immediate vicinity is not above average, and there are several large local employers, the city as a whole has been hard hit in recent years, with unemployment well above the national average Between 5 and 25 per cent of the intake are from ethnic minority familes.

Overview of the school A pilot school within the consortium of local education authorities piloting the
RoA project Oxford Certificate of Educational Achievement (OCEA).

Main areas of activity:

(i)   The school piloted the OCEA P Component with all first year classes in 1985/86, and continued with those classes in 1986/87.

(ii) All first year classes in 1986/87 were involved in a shadowing exercise for the OCEA P component, whereby first year tutors were kept informed about developments within the main pilot tutorial team and invited to introduce these new approaches with their groups, but without specific time or resources to enable this.

(iii) The school also piloted the OCEA Modern Languages G Component with all first year French classes in 1985/86, and continued with those classes in 1986/87.

(iv) Several other areas of relevant work not directly subsumed by the OCEA pilot work but which came to light during the evaluation include:

Several departments were endeavouring to introduce new methods of assessment, recording and reporting.

Work had been carried out to develop and issue a record of achievement for fifth year leavers.

New forms of recording, reviewing and reporting were developed in several other areas of the school such as the DES Special Project (LAPP) and the Alternative Curriculum programme in years 4 and 5.

PRAISE case study Nine visits over eleven days were made to the school for the purpose of
evaluation data-collection between January 1986 and December 1987. The main, but by no means exclusive, focus of study was on the first cohort of pupils to be involved in the OCEA pilot work for the P Component and the Modern Languages G Component, i.e. pupils who were first years in 1985/86. In particular, attention was concentrated upon one class of pupils. This part of the case study comprised: observations of House Periods (weekly tutorial lessons); observations of English and drama lessons (the group's tutor also takes the group for these); observations of French lessons; observations of personal interviews between the tutor and one or more pupils; interviews with a sub-sample of six pupils; and interviews with relevant members of the Modern Languages Department and with the form tutor. Attention has also been concentrated on the work of the OCEA Team of form tutors, through observations of team meetings as well as through discussions with the Senior Teacher in charge of the team's work, and with the school RoA co-ordinator. Discussions have also been held with the head of the Modern Languages department and with the member of the department in charge of the OCEA G Component.

Interviews have been carried out with most heads of department in an effort to document individual departments' efforts to introduce new methods of assessment, recording and reporting.

Regular discussions have been held with the Senior Teacher responsible for the development and implementation of the fifth year summative RoA and the pastoral programme in years 4 and 5 identified under (iv) above.

Regular discussions have also been held with members of staff responsible for the other diverse initiatives identified under (iv) above.

Main issues carried forward for across-site analysis

PRAISE sheets, which were single-page pro-forma attached to 'best' pieces of work selected by pupils with support of their subject teachers and form tutor. The sheets themselves provided space for comments by the pupil, the subject teacher, the form tutor and the pupil's parent or 'friend'. The sheets, and the pieces of work to which they related, were stored in pupils' folders for later reference e.g. when composing end-of-year personal statements. Despite the difficulties encountered in gaining widespread acceptance and co-operation, and in sustaining momentum, it remains an idea worthy of wider dissemination.

The persistent judgement by members of staff in the Modern Languages Department that the G Component assessment framework and pupil self-assessment material was technically flawed and impossible to implement, despite the department's commitment to new forms of teaching and assessment.

The difficulties in sustaining continuous recording in diaries, and the greater success of alternative forms of recording in which the format, purpose and focus changed frequently.

Various strategies to build outward from existing points of strength and/or additional resources within the school, and to incorporate a wide range of disparate pockets of RoA-related activity into a coherent whole-school operation.

DAFFODIL COMPREHENSIVE SCHOOL, WALES
Case study worker: Barry Stierer

School context

Mixed-sex comprehensive secondary school catering for 11 to 18 year olds. Approximately 675 pupils on roll; staff of 41.

Catchment area comprises a large, nearly rural district which also includes some small areas of industry. The surrounding area is economically depressed, with above average unemployment. Most pupils live in private housing.

Overview of the school RoA project

One of eight schools comprising the Gwent Profiles consortium, which in turn constituted one of the pilot schemes within the WJEC RoA pilot project.

All eight Gwent pilot schools piloted the same centrally-developed system.

The main components of the scheme were:

(i) The 'Record of Personal and Social Development': a two-sided card designed for recording personal interests and activities in and out of school and personal qualities. For use in years four and five.

(ii) Subject comment banks developed by panels of teacher representatives from each pilot school.

(iii) The 'Record of Personal Experience': a brief summative statement written by the student in the fifth year based on the 'personal achievements' side of the Record Card ((i) above).

(iv) The Personal and Social Development Summative Comment Bank: 13 comments derived from the 'personal qualities' side of the Record Card.

(v) A Lower School 'Record of Personal and Social Development', based on the equivalent card for years 4 and 5. For use in years 1, 2 and 3.

Progress during 1985/86 was slight, due to teachers' sanctions during the dispute. No recording of achievement took place and no summative documents were issued. No meetings or INSET in school were possible.

Main emphasis in 1986/87 was on producing a summative RoA for all fifth year pupils by early April 1987. Recording of personal achievements and qualities were gradually introduced in years 1 to 4.

PRAISE case study evaluation

Fieldwork was not able to begin until the autumn of 1986. Six visits were made to the school for the purposes of data-collection between October 1986 and November 1987. The main foci of study during 1986/87 were:

The use of the 'Record of Personal and Social Development': observations of tutor-pupil discussions in the fourth year and interviews with form tutors and pupils in years 4 and 5;

The work of the subject panels, through interviews with those members of staff who participated as representatives of the school;

The production of a summative RoA for fifth year pupils in 1987, through interviews with pupils, tutors and subject staff.

At the same time, data were collected relating to the other strands of activity in the school, largely through interviews with heads of year, heads of school, heads of department and the school RoA co-ordinator.

The direct recording of personal qualities, and implications of that recording for formative processes and teacher-pupil talk.

The central consortium development of all RoA elements, based on a comment bank format, and the uniform implementation of those elements in all pilot schools.

Model of development based on (a) management objectives which were, compared with other schemes, relatively utilitarian amd summative-orientated and (b) a formative process in years 1 to 4 derived from the project's summative starting point.

The adoption by the Mathematics Panel of a staged assessment model for the design of its comment bank.

The practical difficulties at the early stages of integrating tutor-pupil interviews with existing work load in the absence of earmarked time.

Main issues carried forward for across-site analysis

DAISY HIGH SCHOOL, WALES
Case study worker: Barry Stierer

School context

Mixed-sex comprehensive secondary school catering for 11 to 18 year olds near the centre of a large Welsh city. Approximately 950 pupils on roll, staff of 60.

Catchment area covered a range of housing, although the majority of pupils came from the council estates in the district. Fewer than 5 per cent of pupils came from ethnic minority families. The local area was one of economic decline, characterised by above-average unemployment. There were no large employers in the immediate local area, although several were located in the city centre.

Overview of the school RoA project

One of two schools in the county taking part in the WJEC RoA pilot project. Although the co-ordinators of the two pilot schools in the county conferred regularly, the two schools developed their schemes essentially autonomously. The common reference point was the WJEC criteria rather than a distinctive LEA approach.

Progress during 1985/86 was slight, due to industrial action. No recording of achievement took place and no summative documents were issued. No meetings or INSET were possible. Main activity was by the school co-ordinator in promoting awareness of the project on a one-to-one basis with e.g. heads of department, and encouraging the development of departmental comment banks.

Main emphasis in 1986/87 was on producing a summative RoA for all fifth year pupils by early April 1987. At the same time, recording of personal achievements, and interviews between pupils and form tutors, went on in the fourth year and a personal recording system was gradually introduced to tutor groups in years 1, 2 and 3. Discussions were held with heads of local junior schools regarding the development of a primary school RoA.

PRAISE case study evaluation

Fieldwork was not able to begin until the autumn of 1986. The main focus of study during 1986/87 was on the fourth year, and in particular on a single fourth year tutor group. This took the form of observations of tutor-pupil interviews as well as interviews with the form tutor, with pupils in the group, and with a number of other fourth form tutors. At the same time, data were collected relating to the other strands of activity in the school, largely through interviews with Heads of Department, with form tutors in Year 5, with the Head of Lower School, with form tutors in the Lower School, with the Head of a local junior school, with the school RoA co-ordinator, and with the headteacher. Nine visits were made to the school for the purposes of data-collection between July 1986 — December 1987, though most evidence in this report relates to work in the school year 1986/87.

Main issues carried forward for across-site analysis:

The recording of personal qualities using a comment bank approach.

The procedure adopted for the collation and aggregation of different teachers' contributions to the summative 'Personality and Attitude' statement, whereby a panel comprising of several senior members of staff and a form tutor examined each comment to go forward from the comment bank to the final document.

The element of conflict between the specific recording purposes of the one-to-one tutor-pupil interviews and the wider (and more time-consuming) pastoral uses to which some tutors put the interviews.

The envisaged function of the 'Personality and Attitude' comment bank (personal qualities) of arresting the perceived decline in common courtesy and self-discipline among pupils.

The development of a junior school RoA in collaboration with local junior schools.

The constraints imposed on departmental development of subject comment banks by the limitations of the in-house computer programme.

The validation function of the WJEC criteria.

The uniform adoption of a comment bank approach for the recording of subject-specific achievement.

HONEYSUCKLE SCHOOL, WALES
Case study worker: Susan McMeeking

School context   Honeysuckle School was an 11 to 16 mixed comprehensive school situated on the outskirts of one of Clwyd's major towns. Opening as a secondary modern in 1958, the school expanded rapidly following reorganisation in 1972 and comprehensivisation in 1973.

In September 1986, there were 54 members of staff and 902 pupils although the school roll was falling. It was anticipated that the school roll would drop to approximately 800 pupils by 1989. The percentage of ethnic minority pupils in the school was low at less than 5 per cent.

At the time of writing the school was drawing increasing numbers of pupils from private housing but over 50 per cent were still drawn from council housing and many pupils were described as coming from deprived areas. Over 50 per cent of the pupils were bussed in from the outskirts of the town, areas still regarded as 'villages' by the local inhabitants. Each 'village' was described as having a distinct character, for example, in one, life was centred around its local steelworks and was described as a 'born-work-die' community; another, an overspill village for the town was more 'sophisticated', pupils from this area seeking out social life in the town. The majority of the school population was drawn from these two communities.

The area was one of economic decline with an unemployment level of 18 per cent. There had been a decline in the traditional steel and coal industries but there were still a number of large employers in the area. These industries were based on industrial estates, many of which employed predominantly female workforces and consequently boys from Honeysuckle found it more difficult to find work than the girls. Further education was not traditionally attractive to the pupils at the school but was becoming increasingly more popular. A breakdown of the destinations of the 1985/86 school leavers revealed that 35 per cent went on YTS courses, approximately 25 per cent of pupils went on to the local sixth form college and into other FE institutions, 25 per cent found employment (more girls than boys), and 15 per cent were unemployed, had moved out of the area, were sick or generally unavailable.

Overview of the school   During the period of the evaluation, Honeysuckle continued to implement the
RoA project   Clwyd RoA scheme. The school had been using this scheme since its creation in 1982 although it went through a number of drafts. Described as a cross-curricular scheme, it operated in the fourth and fifth year during which assessments were made of pupils' numeracy skills, personal skills (i.e. practical skills, oral skills, creative skills, physical skills, and literacy), assessed by staff if they felt the skills were relevant to their subject areas, and personal qualities (response to others, response to work in school, response to activities in school), assessed by all staff in contact with a particular pupil. These assessments were discussed in reviews between pupils and their form tutors.

Assessment of personal skills and personal qualities formed part of a summative document issued for fifth year leavers. Other information in the summative document included, figures for attendance and punctuality, courses followed during the fourth and fifth year, a record of pupils' interests and achievement, information on pupils' school and community service, and details of pupils' work experience undertaken during their fifth year at Honeysuckle. An overall comment prepared by the form tutor fronted the profile.

PRAISE case study   Fieldwork began in November 1985. A fourth year form group was selected for
evaluation   close focus study during 1985/86 and 1986/87 after which they left the school. A second fourth year group was selected for study during 1986/87 and 1987/88.

Observations of review sessions between both form tutors concerned and their pupils and follow-up interviews with tutors and pupils were carried out.

Interviews were carried out with the school co-ordinator, the headmaster, the deputy in charge of the curriculum, heads of department, other form tutors and pupils involved in the scheme, and non-involved staff.

Questionnaires were sent to the parents of pupils in the sample and to employers of school leavers.

The use of negative comments in the summative document.

The co-ordination of academic staff assessments by pastoral staff.

The anonymous nature of these assessments and the difficulties this presented for effective diagnosis and target-setting.

The scheme's lack of impact on the curriculum.

Main issues carried forward for across-site analysis

YSGOL EITHIN, WALES
Case study worker: Colin Morgan

School context

Ysgol Eithin is an 11 to 18 four form entry comprehensive school situated in a small market town (population 1900) in the heart of rural Wales with a catchment area of some twenty five miles at its widest and receiving children from fourteen primary schools. It sends pupils to eight colleges of further education.

Agriculture and related services, tourism, and 'boutique industries' constitute the economy of the catchment area. Tourism and services are on the increase and unemployment appears to be about the Welsh average, though in the perception of local *cognoscenti* 'there is little real unemployment'.

This area is not one of marked differences in the socio-economic status of its people to the degree that would be true of Britain as a whole and certainly not in terms of the south of England. The farms are small family farms.

Many families do not own a car, and more than a third of the pupils live in council-owned houses. Social homeogenity rather than hetereogenity is what strikes the city visitor from South Wales. Indeed, what stands out more than the average is the marked disposition of the area for social, cultural and community involvement; competence in musical and dramatic activity appearing to be a special and long-established local talent.

Welsh is the **main language** of the school's catchment area. At the 1981 census over 80 per cent of the catchment population were Welsh-speaking, and many of the subjects taught in Ysgol Eithin are taught through the medium of Welsh. In fact, only four of the thirty eight staff at the time of our visits were non-Welsh speaking though some of these were form tutors taking part in the RoA experiment.

When we interviewed the pupils on an individual basis, we gave them a choice regarding which language they wished to use. Almost threequarters (74 per cent) chose to speak with us in Welsh, a graphic indicator to outsiders, as we were, of the strength of the Welsh language throughout this community despite the crushing supremacy of English in modern mass media, the effects of 'good life' English settlers, and the possible effects of second homes in the area. The confidence to choose Welsh with visitors from far outside their own locality was greater with the boys (79 per cent) than with the girls (66 per cent). We think we have noticed this elsewhere in our school evaluation work in the Welsh-speaking areas but do not know why it is so. Indeed, an issue arising from our evaluation of the RoA experiment at Ysgol Eithin concerns the language question. The formative part of the RoA Ysgol Eithin has pioneered represents major investment in group and one-to-one tutorials. Hence language and social interaction skills and language are crucial to the full success of these elements. Pupils are not neutral about which language is used for these activities.

Overview of the school RoA project

The RoA experiment at Ysgol Eithin began with the four fourth year tutor groups in September 1985, and the first RoA (Profile) documents were produced and distributed to these pupils in July 1987. The RoA document provided each pupil with a profile of judgements based on comment bank options. All departments except maths had completed coding sheets. These could be input by light pen into a computerised record which, after amalgamation and a degree of negotiation for pupil agreement, formed the final document. Pupils were to receive the record of achievement documents at a special ceremony before the end of the fifth year to which their parents were invited. The ceremony was held to hand over the attractively printed document covers but their contents had not then been finalised and they were handed back at the end of the ceremony for completion in time for pupils to take them home at the end of the term.

It is important to record that the development of the experiment at Ysgol Eithin faced the same set of wider difficulties confronted by schools elsewhere during the period of this radical innovation. During the two year period of start-up and experiment to be assessed as part of the England and Wales independent evaluation by The University of Bristol and the Open University of records of achievement, the school did not escape the sanctions and disruptions of the continuing teachers disputes with their employers and the government.

The main elements of the records of achievement experimented at Ysgol Eithin have been:

(i) A designated co-ordinator to lead the experiment and a team of four tutors for each year group to implement the agreed elements of the experiment;

(ii) Group tutorial work once a week for twenty minutes;

(iii) One-to-one tutorials each term over two years;

(iv) Pupil files holding key written work from (i) and (ii), the emerging profile and other inputs to it. These are owned by the pupils and kept accessible to them; and,

(v) A Profile leaving document of 'achievements' chosen from a distance scale of four comments for each of the following: Personal Qualities; Oral Communication Skills; Language across The Curriculum; Written Communications Skills; and Practical Skills.

It seems that the school's involvement in the RoA experiment came about by the Headteacher being given the proposition at an in-service day by the LEA Chief Advisor. In short, the school was 'drafted' to be one of the LEAs two pilot schools, and the (then) deputy head made responsible as co-ordinator. In the second year of the project the Headteacher was seconded to the Inspectorate so that his Senior Deputy who was also the RoA co-ordinator became the acting Head, and the role of co-ordinator was taken over by one of the participating tutors — a head of department. Prior to the advent of the pilot scheme there had been no development of profiling nor of active tutorial work on a group or individual basis. Whilst there was also no great formalisation of pastoral roles or policy in the way that there has been in many schools, the school was of a size to lend a family ethos, and the staff perspective was that they did in fact know the children very well. Nevertheless, under these conventional arrangements teacher-pupil contact on a one-to-one basis usually meant disciplining miscreants or more rarely, and then with a minority, the praising of certain high achievers for their notable known success in a school activity.

Regarding the start-up at Ysgol Eithin induction and briefing was done by experienced visitors from the WJEC and a neighbouring LEA with the holding of two formal INSET days. After that the key participants at Ysgol Eithin perceived that they trained themselves and ploughed their own furrow in developing their own scheme through the co-ordinator and tutors meeting as a group. They did not take a scheme off the shelf, neither did they co-ordinate what they were to do with the LEA's other RoA school.

Neither employers nor parents were part of the development work, nor had a position regarding accreditation been determined during the period of evaluation though the Ysgol Eithin scheme was being reported to and reviewed by the WJEC who during this period were actively formulating a policy on accreditation and profiling generally. There were, though, periodic meetings of a committee convened by the LEA profiling co-ordinator but from our examination of some of the minutes, we saw these events to be concerned with information exchange and discussion of key ideas connected with records of achievement rather than to drive a countrywide policy. The school believes they have been autonomous in their RoA developments and the evaluation evidence would wholly support this contention. Prior to the RoA experiment the policy for reviewing, recording and reporting pupil progress was by means of bi-termly assessments and biannual reports, and this continued in the period the experiment was under evaluation.

PRAISE case study evaluation[1]

Specifically, the data on which this report is based are: notes made whilst observing Dr. Patricia Broadfoot interviewing staff and pupils on an initial visit to the school in 1986; analysis of relevant documents provided for us by the school; sixteen interviews with school staff (i.e. RoA tutors, the co-ordinator, and senior management); thirty four interviews including self-completion exercises with pupils from cohorts one and two of the experiment held in May, June, October, and December 1987. In addition, observations were made of group tutorial sessions but not of the one-to-one sessions.

It should be noted that we did not construct the samples of pupils to be interviewed because the necessary base data (particularly pupil ability category) on which to create stratified samples could not be made available to us. Instead we requested the school prior to our visits to the fourth and fifth year cohorts to provide us with a cross section of pupils from the year groups. We had no way of knowing whether, regarding ability or courses followed, the pupils we interviewed were in fact a representative sample of the range in the year group.

---

[1]This report derives from visits to the School and the main receiving College of FE to collect the data between November 1986 and December 1987 by Colin Morgan and Janine Davies. The evaluation design, data analysis, and this report were prepared by Colin Morgan.

An outline of methods mainly in relation to the case study element of the evaluation

Introduction and rationale

At the beginning of this report we made it clear that the greatest part of our time, effort and resources was devoted to studies of the development and implementation of records of achievement systems in individual institutions. Our reason for choosing this particular approach from a range of possibilities was based on our assumption that progress and results of the pilot initiatives **at school level**, where they impinge on the experience of pupils, was of the utmost importance. We also knew that the experience of recording achievement at school level was likely to be intimately related to complex and inter-related features of each school's context. We wanted, therefore to adopt a methodology which would preserve the sense of uniqueness of development in particular contexts whilst also giving us purchase on common themes and issues. For this reason an approach through multi-site case study seemed to offer the best hope for achieving this goal. There were precedents for this[1] and we were particularly attracted by the model of the Cambridge Accountability Project which built in an interim reporting stage as a means of refining the issues of principal concern to both researchers and participants. Although, unlike that project, we did not see ourselves as having a prominent formative role in relation to projects, we perceived that a similar strategy would enable both progressive focusing and respondent validation in relation to our analysis of the enormous range and variety of themes associated with recording achievement in schools.

Selection of schools for case study work

At the beginning of our evaluation, a decision was taken to select 24 schools for case study work. It was recognised that this was an enormous load for the three full-time researchers on the team who were expected to conduct seven or eight case studies each, but it was felt important to select two schools in each single-authority scheme, one school in each of the local authorities participating in the the English consortia, and one school from four of the eight local authorities in Wales. Schools were not chosen, however, to 'represent' the salilent features of the scheme in which they were taking part. Indeed we were sceptical of the notion of a 'typical' school as far as any scheme was concerned. Instead we sought a kind of 'representativeness' on a national, rather than local, scale in terms of features of larger populations or groups of schools generally. To this end we selected schools in relation to a number of dimensions which we had grounds for suspecting might influence or explain differences in records of achievement development and implementation. In particular, we wanted to ensure that our final selection provided us with an opportunity to study the possible influence of each of the following features in at least one setting:

Catchment area dimensions:

Inner city, suburban, town and rural areas

Areas of economic growth, stability and decline

Council housing or private housing

Above average, average, or below average local unemployment

Variable proportions of pupils from ethnic minority families

School organization dimensions:

Middle school, upper school, 11 to 16 secondary, 11 to 18 secondary, tertiary college

Special school

---

[1]Ebbutt, D. 'Multi-site Case Study: some recent practice and the problems of generalisation', *Cambridge Journal of Education*, Vol. 18, No. 3, (1988).

Voluntary-aided school

Comprehensive and selective schools (grammar and secondary modern)

Large and small schools

Welsh-speaking or bilingual school

Specific circumstances relevant to records of achievement:

Previous experience and no-previous experience of profiling or RoA

Related curriculum developments (curriculum review, pastoral programmes etc)

Involvement with other related initiatives (CPVE, etc)

Target groups ranging from year 1 to year 5.

Basic data were collected from local co-ordinators in relation to these dimensions and a provisional list of 12 schools was drawn up which together satisfied all the criteria inherent in these dimensions. It should be noted, however, that the choice on some dimensions was very limited because, for example, the schemes funded by the DES tended not to be in LEAs with large numbers of inner city schools with high proportions of pupils from ethnic minority families or high rates of local unemployment. This feature was compounded by the fact that inner city schools tended not to be included in the pilot even in LEAs which contained significant numbers of them. With the exception of one local authority (Lincolnshire), schemes invited schools to bid for inclusion in the pilot. 1985 was a difficult time in schools and most bids came from schools which did not have to contend with the worst problems associated with inner city environments.

Once these 12 schools had been selected, a second group of 12 were chosen effectively at random, although such factors as accessibility via road and rail routes and the recommendations of local co-ordinators were taken into account.

Finally, consultations took place with both HMI and local evaluators to ensure that our selection of schools did not overlap with samples of schools chosen for in-depth study by other groups. One scheme co-ordinator had identified seven levels of evaluation (ranging from national to in-school) in relation to this initiative and the dangers of over-evaluation were very real.

At this point we should note that in the course of our case study work, two schools in our original selection were 'lost'. For reasons principally to do with difficulties arising from teachers' union action, evaluation work in one school was never started whilst in the other the data collected were so slight that the idea of producing a case study report for publication was abandoned. In both cases this was unfortunate because one school represented our only example of a boys' secondary modern school whilst the other was an inner city school.

Gaining access

Access to schools was negotiated first with local co-ordinators, then with schools, on the basis of two documents. One was a site brief (see Document 1) which introduced the case study element of our evaluation and outlined both the kinds of data we wished to collect and the reports we hoped to prepare during the three year funding period. The second (see Document 2) set out the ethical procedures we intended to adopt in order to open the processes within particular schools to public scrutiny whilst at the same time giving individuals within these settings some control over, and therefore confidence in, the reporting process. The balance is notoriously difficult to achieve and has been a principle concern of educational evaluators working within a tradition of 'democratic evaluation'. In a number of cases ethical procedures have been derived from the principles of 'accessibility', 'negotiation' and 'confidentiality'

APPENDIX 2
An outline of methods mainly in relation to the case study
element of the evaluation

(MacDonald and Walker, 1977[1]; Simons, 1979[2]; and Kemmis and Robottom, 1981[3]) and our ethical guidelines were developed with these exemplars very much in mind. In this respect it is worth pointing out that the principles and procedures we adopted in our evaluation mirror many of the principles and procedures underlying approaches to records of achievement for many of the same reasons. Whilst our formative role in relation to participating schools was limited we were concerned that public reports should be agreed with the schools on the basis of negotiation, thus giving them some sense of ownership.

As we now report the results of our evaluation we feel reasonably confident that this strategy for gaining access to the work of schools was moderately successful because, of the 24 schools we originally approached, we now have publicly available accounts of 22.

Data collection

The site brief, like our ethical guidelines, held up reasonably well in execution and we were able to collect most of the kinds of data for which we had hoped. Our decision to focus on the experience of a small sub-sample of pupils in each school was particularly helpful in that it provided an organizational focus for many of our visits, around which we could fit interviews with staff. Occasionally, we drew criticism for focusing on such a small group within each school, but had we not done so we believe we would have found it very difficult to observe activities such as review sessions or to have gained any coherent picture of the experience of individual pupils. In all schools some pupil data were collected in relation to larger groups and over our total sample of 22 schools our data on processes in relation to individual pupils were quite substantial.

There were however a number of respects in which we over-estimated what it was possible to accomplish within our resources. First, our estimation of 30 days of data collection in each school proved to be unrealistic and, on average, nearer half this number of days was spent on each site during what amounted to little more than two years for fieldwork. We found, however, that the number of days in the field did not always correspond to the quantity or quality of data collected. One of us, for instance, discovered that she had made most visits to a school where she had difficulty in gaining access to relevant information. On the other hand, her less frequent visits to another school were always packed with activity which gave her rich data.

Secondly, we were advised by experts in the field of psychometrics to abandon any great hope of using inventories to assess whether increases in pupil self-concept of self-esteem had occurred as a result of involvement in records of achievement processes. No 'off-the-shelf' instrument was available for this purpose and the development of a new instrument, or the re-validation of an existing one in the context we had in mind, would have taken all the time and resources available to us. We also found it difficult to isolate causal factors within the complex social settings in which RoA systems operate. We regretted the loss of this kind of evidence as an element of case data but we felt that we had to give priority to the information needed to guide policy making. Whereas statistical data of this kind might tell us whether recording systems had achieved one of their stated goals, its capacity to tell us why, or to illuminate the policy alternatives, would have been very limited. Our decision, therefore, was to draw on pupil, teacher, and sometimes parent, perceptions, collected in a somewhat less systematic way, for qualitative evidence of changes in motivation.

---

[1]MacDonald, B. and Walker, R. 'Case-study and the social philosophy of educational research' in Hamilton, D. et al (eds) *Beyond the Numbers Game*, Basingstoke and London, Macmillan Education (1977).

[2]Simons, H. 'Suggestions for school self-evaluation based on democratic principles, *Classroom Action Research Network, Bulletin No. 3*, Cambridge Institute of Education (1979).

[3]Kemmis, S. and Robottom, I. 'Principles and procedures in curriculum evaluation', *Journal of Curriculum Studies*, Vol. 13, No. 2 (1981).

APPENDIX 2
**An outline of methods mainly in relation to the case study
element of the evaluation**

Analysis

In relation to all our data (including the documents which supplied the data for the meta-evaluation strand of our evaluation, reported in Part Two) we defined our analytical task as eliciting the kind of information that the national steering committee (RANSC) would need in order to draw up draft national guidelines. Thus our prime responsibility was to illuminate practice in such a way as to assist in the process of policy making, mainly at national level although we hoped that local and school-level decision makers would also find our analysis useful. We were less concerned to make judgements in terms of some overall criterion of success, or indeed to judge individual developments in terms of their own specific objectives, as to illuminate the themes and issues that future policy would need to address.

As a starting point we developed a framework for analysis derived from the criteria and issues already identified in the 1984 statement of policy (Document 3 is the briefest of several formulations used by the PRAISE Team in its analytical work). As work progressed, other themes and issues emerged and were added to this list. For instance, the transition from the formative to the summative stage in recording, and the nature, status and use of interim summary documents came to have some prominence, as did the issue of equal opportunities and questions arising from the management of the initiative at scheme, LEA and school-level.

This framework for analysis was progressively developed and refined in an iterative process of dialogue with participants in the evaluation. The main vehicle for this dialogue was the series of reports produced by the team.

Reflections on the case

At this point, because our evaluation methods focus largely on case study, it is probably worth adding a note about our conception of the bounds of the case. For the most part, when we talked of 'the case', we were referring to development and implementation in relation to the records of achievement pilot projects within individual schools between autumn 1985 and autumn 1987. This, itself, gave us some headaches since we often found developments in schools that were conceptually or structurally linked to records of achievement but were not perceived to be part of the pilot initiative. Thus we were faced with decisions about whether or not we should investigate these initiatives as well as those that were clearly ESG-supported. Often we had to use our discretion in the light of the special circumstances of each context. We usually took related developments into consideration although we continued to concentrate our main efforts on developments specifically associated with the pilot scheme in the school.

On a different level, however, there emerged a sense in which it was not the schools, but the whole collection of case studies that were **the case**. In this respect, Sadler (1981) conveys an important insight in relation to the kind of multi-site evaluation project which we were engaged in: 'The art of assimilation (generalisation) is to see the multiple site investigation not as a collection of case studies but as **the case**. The natural unit for each fieldworker (the school or the institution for example) has to be seen less as an entity, more as an element' (p.21)[1]. For PRAISE, therefore, the case was really the operation of the DES-funded initiative rather than records of achievement in X school or Y scheme, although what happened in X school or Y scheme was crucial for any understanding of the total case.

1986 interim report

In the summer of 1986, we produced a first interim report on emerging issues derived mainly from brief non-puplic accounts of the main issues emerging in each school as perceived by case study workers. No particular strucuture was prescribed for these single-site statements of issues. The across-site analysis

[1]Sadler, D.R. *Basing evaluation and policy analysis on multiple case studies: some considerations of utility, sampling and assimilation*, Department of Education, University of Queensland (mimeo) (1981).

231

necessary to produce the interim report of the case study strand of our evaluation, together with a similar analysis of interim reports from scheme directors, provided an agenda for subsequent analytical work which covered both substantive RoA themes and management issues.

1987 interim report

In the summer of 1987, a second, though much more substantial, interim report was written. In this instance, however, the means of production was much more elaborate. For the across-scheme analysis of project directors' reports the headings derived in the previous reporting stage were used in a revised form. The headings in the previous across-site analysis of school case studies were also used as a starting point, but on this occasion each of the three full-time case study workers independently attempted a preliminary across-site analysis of the seven studies for which they were each responsible, identifying illustrative material where this was appropriate. The resulting notes were then discussed at a residential team meeting in Cambridge with the result that individuals were given responsibility for writing sections of the report to be submitted to RANSC based on fairly detailed guidance from the team concerning content and structure. With regard to the latter it was decided that headings should identify key topics and that these should be mapped out to clarify the nature of each topic, to outline the range of approaches observed, and to set out the different ways in which terms were understood. This mapping exercise was followed by a number of relevant examples drawn from all the case study evidence to illustrate the diversity of approach and viewpoint in the context of each topic. On the whole we chose to include discrete examples in order to preserve our qualitative approach and to avoid any potentially invalid and unreliable use of quasi-statistics (e.g. 'some schools . . .', 'a few schools . . .' etc.). Finally, we closed each sub-section of our report with a number of questions or issues which we felt needed to be given special consideration in relation to the particular topic.

Whilst this interim report was in draft form it was sent both to local and school co-ordinators who were invited to comment on the factual accuracy of descriptions, the fairness of interpretations, and the relevance of the issues raised, especially in relation to examples drawn from their work. All references were in the nature of general, 'anonymised' summaries so we judged, in accordance with our ethical guidelines, that no special clearance was needed. However, we endeavoured to take account of responses in our final draft as far as we felt reasonably able to do so. For the most part, responses to specific references were relatively few and easily dealt with. Other responses, mainly from local co-ordinators who had perhaps the greatest stake in the developments, were of a general nature concerning the overall style and format of our report. There was clearly some anxiety on four counts: our decision to make the report issues-focused; the extent to which we engaged in interpretation of our data and the degree of influence this might exert on the deliberations of RANSC; our decision to select evidence from schools as illustrative of particular approaches to specific themes rather than to demonstrate our grasp of the holistic qualities of developments in individual contexts; and the possibility that illustrations from particular schools would be taken by readers to represent development in the scheme as a whole. We recognised that these were serious points and we consequently included in our 1987 interim evaluation report a note on the status of our across-site report as we perceived it. We have reproduced our response to these points in this final report (see Document 4) because the way we have chosen to structure our present report is broadly similar and, therefore, some of the points raised in relation to our 1987 interim report may still apply.

This final report

Our decision to continue to develop this approach for the purposes of this final report was based largely on the gratifyingly positive response our interim report received once it became public. Not only did the structure appear to be useful but we were led to believe that we had identified most of the main themes and issues relevant to the policy-making area. There were however areas where our data and analysis were still thin (e.g. INSET, resources, and pupil differences) and we identified these as areas for special attention in the last phase of data collection.

In a number of important respects, however, our preparation of this final report was substantially different from the previous two. Most importantly we conducted our across-site analysis on the basis of complete case study reports of each of our 22 case study institutions. As our original site brief (Document 1) indicated, we had hoped to prepare interim case study reports prior to writing our interim across-site report in 1987. In January 1987, we began such an exercise but at that time we had not developed a clear structure for such reports and tended to produce lengthy chronological accounts of development. As time was pressing we abandoned the task of interim case study report writing and decided to move straight to the preparation of the interim across-site report based on the strategy described above.

Once this task was complete we turned our attention once more to the preparation of case study reports. We felt under a strong obligation to give something back to the schools with which we had worked, but we also felt that reasonably comprehensive case study reports of the project in individual schools would help us to validate the analysis in our interim across-site report and ensure that illustrations selected for the final report were the most appropriate. In order to assist us in the task of qualitative comparison which we would face in our final across-site analysis, whilst still preserving a sense of the uniqueness of individual developments in particular contexts, we chose to structure every case study report on the basis of the themes and issues identified in our interim across-site report. In other words, although we tried to describe development in relation to each theme or issue, our case study reports were thematic rather than chronological. This did not always make for smooth story-telling because there was no obvious beginning, middle and end, and some information had to be repeated under several headings because many themes were interrelated. However, reports structured in this way created a useful second-order data-base for our across-site analysis and enabled us both to scrutinise all cases in relation to the same themes and to lift particular discrete illustrations straight into Part One of this report.

When draft case study reports were to hand, we once again began the process of across-site analysis. This largely followed the pattern of the previous year although our task was made easier in some respects by the existence of our interim report. To begin with we examined the headings and mapping sections in our previous report in the light of our completed case studies and revised these where appropriate. We also selected extracts from our case study reports to illustrate, in some instances, the range of experience within schools, in others, the special experience of perhaps only one school in relation to a particular point. Although relevance of the illustration was our main criterion, we tried to be even-handed with schools and ensured that most schools had between six and 12 references made to their work.

In terms of qualitative methodology this form of analysis came close to the 'constant comparative method' described by Glaser and Strauss (1967).[1] Hammersley and Atkinson (1983) summarise this in the following way: 'Each segment of data is taken in turn, and, its relevance to one or more categories having been noted, it is compared with other segments of data similarly categorised. In this way, the range and variation of any given category can be mapped in the data, and such patterns plotted in relation to other categories' (p.180)[2].

---

[1]Glaser, B. and Strauss, A. *The Discovery of Grounded Theory*, Chicago, Aldine (1967).

[2]Hammersley, M. and Atkinson, P. *Ethnography: principles in practice*, London, Tavistock Publications (1983).

Finally, we looked again at the issues which we had raised in our interim report and made some decisions about the extent to which our evidence gave us justification for coming to some conclusion or judgement in relation to the theme, or whether the issue appeared still to be unresolved. At the outset of our project, as our ethical guidelines (Document 2) indicated, we had stated our intention that our reports should contain description, interpretation and judgement. It was also made clear to us that RANSC expected an element of judgement in our reports. However, we decided to stop short of explicit recommendation believing this to be the proper role of RANSC in its own final report to the Secretaries of State.

In the case of this final across-site analysis we did not seek clearance for the **whole** report from participants in our case studies. Instead we went through an elaborate clearance procedure in relation to our individual case study reports by asking participants to comment on any factual inaccuracy or possibly unfair interpretation. First we cleared specific references to the statements or work of individuals who, although referred to by role designations, might be recognised when our accounts became public. Having revised the drafts in the light of their responses, complete case study reports were then sent to schools for further comment. Finally when accounts had been cleared at school-level they then went to LEA and scheme co-ordinators who were also invited to respond. It would be foolish to pretend that this was in all cases an unproblematic process or that we were able to satisfy the worries of all individuals or groups on all counts. Indeed in some instances responses simply underlined differences in perception rather than secured agreement. There was concern in some quarters about the fact that our data-collection period ended in the autumn term 1987 and consequently could not take into account refinements which schools and schemes implemented at the very end of the three-year pilot period. An anxiety at LEA level about the representativeness of schools in relation to the scheme remained strong, as did the tendency of the issues-focused approach to highlight problems rather than successes. In these cases, short of writing a completely different report, which we were unwilling to do, we felt that we could do little more than reiterate the points we had made in relation to the status of our 1987 interim report (see Document 4). However, we decided that the short introduction sections from each case study report should be included (see Appendix 1) in order to give readers a sense of the totality of the project in each case study school and to elaborate special features of context which might make it atypical of the scheme with which it was associated.

Having gone through this clearance procedure, and revised drafts where necessary, we then considered case study reports to be 'in the public domain' and available, both for the selection of illustrations for this report, and for future publication in their entirety. Insofar as the second order data-base of case study reports had been cleared as accurate and fair, we saw no need to clear our across-site analysis as such. Whilst we accept that some people might disagree with our interpretation of our evidence (which will be available as a collection of case study reports in mimeograph form from our universities), we feel that we have a right to make our own interpretation and, as is the convention with academic research, any further debate can take place in the public forum.

Validation   Qualitative case study methods tend to be strong on validity if short on reliability and replicability. As we mentioned earlier, we saw our clearance procedures, not merely as an ethical means of giving those who are made most vulnerable by the evaluation some measure of control over what is reported, but also as a way of validating the accuracy of our descriptions and the plausibility of our interpretations. In addition to this 'respondent validation', however, we also used a **read and react exercise** in relation to our 1987 interim report as a means of cross-validating our analysis against experience outside our case study schools. This asked staff within a sample of schools to respond to the content of our across-site analysis of case studies with special reference to their perceptions of the extent to which their experience was in accord with our findings.

For the purposes of this 'read and react' exercise, which had taken the place of our originally planned teachers' survey, a sample of sixty schools was chosen from a population consisting of all the schools listed by co-ordinators of the nine ESG-funded schemes and the nine non-ESG schemes with which RANSC also had contact (i.e. West Midlands Records of Achievement, Springline Trust, Cambridge Partnership for Records of Achievement, Gulbenkian, South Western Profile Assessment Research Project, South East Records of Achievement, Pupils' Personal Records, Wiltshire Records of Achievement and Profiling Project, Northern Partnership of Records of Achievement). Some schools were, however, excluded from the sample at this point if they were already being studied by PRAISE or HMI, or if, for some other specified reason, their inclusion was considered inappropriate. In order to ensure that all eighteen schemes were represented, a sub-sample was drawn at random for each scheme. From the larger schemes, four schools were included; three from smaller schemes; and two or one from the smallest schemes. Allocation of schemes to each category was to some extent arbitrary, being mainly determined by the need to achieve a final total of approximately sixty.

Once the sample had been drawn up a copy of our 1987 interim report was sent to the headteacher of each school requesting comments on the issues in the across-site section in particular. We were especially interested in any judgement on the extent to which it was possible to generalise our findings to experience of profiling and records of achievement within these schools. The response was disappointing to the extent that few schools commented on the detail of our report, and we had little evidence that it had been circulated among staff. Instead, and perhaps understandably, many schools sent us general materials from their own projects. Whilst these were interesting, and added to our growing archive of records of achievement materials, they did not provide the kind of evidence that might have encouraged us to revise our approach or our framework for analysis. The fact that so little of our analysis was challenged had to be taken as some confirmation that we were at least on the right lines.

Another strategy for validating our analysis derived from the relationship of our work to a concurrent exercise carried out by HMI, which was scheduled to report at approximately the same time. As we mentioned in relation to our selection of schools for case study work, we consulted with HMI in order to avoid working in the same schools. We also agreed to share our draft final reports, prior to publication, in order to provide a form of cross-validation. This was a mutually useful exercise and we were able to conclude that whilst approaches, presentation and emphases sometimes varied, the substance of the two reports were remarkably similar.

Finally, an internal form of validation was provided by the two major strands of our evaluation which are presented in Parts One and Two of this report. To a certain extent these were worked on independently by different members of the PRAISE Team, and the analysis in one can be set against the analysis in the other. Although the different foci of these two parts, as well as the different sources of data, have contributed to some different emphases, there nevertheless appeared to be sufficient commonality of insight for the synthesis in Part Three.

In the end, however, our analysis of progress and results within the pilot schemes will be validated if our observations and conclusions strike a chord with the experience of our readers. In the jargon, this is known as 'naturalistic generalisation' and is perhaps the form of validation that really counts.

 PILOT RECORDS OF ACHIEVEMENT IN SCHOOLS EVALUATION

THE OPEN UNIVERSITY, SCHOOL OF EDUCATION, WALTON HALL, MILTON KEYNES, MK7 6AA
Telephone (0908) 653753 (Direct line)

UNIVERSITY OF BRISTOL, SCHOOL OF EDUCATION, 22 BERKELEY SQUARE, BRISTOL, BS8 1HP
Telephone (0272) 24161 (Ext M371)

Please reply to: Ref: 405010U

AN OUTLINE PROPOSAL FOR SCHOOL CASE STUDIES

Introduction

Researchers from Bristol University and the Open University plan to make 23 or 24 case studies of individual schools within the DES-funded Records of Achievement Pilot Schemes. This will form one important strand of their independent evaluation of the pilot schemes.

Case studies will be based on the equivalent of ten days fieldwork, per year, in any one school (giving a maximum of 30 days over the three years). The following is a summary of what the researchers hope to accomplish in this time, if circumstances permit. School teams are asked to consider this outline at the outset of the research and offer any comments or queries to the researcher involved with their school.

Case study aims and methods

As a basic minimum, the case study evaluation aims to illuminate the purposes and issues, identified in the DES Statement of Policy, as they emerge in individual schools.

To this end the PRAISE researchers propose to carry out the following studies in each school. The details will be discussed with staff who are likely to have ideas about what is most relevant in their own case.

1. A general school study

It is envisaged that information will be gathered from:

a) Observation of some planning meetings, e.g. school team (some with LEA coordinator), staff meetings;

b) Observation of some meetings/events when plans/progress are communicated to pupils, parents, employers/community;

c) Collection of relevant school documents (internal and for public consumption);

d) Interviews with senior management, some or all of the school project team, and a small sample of non-involved teachers (if any);

e) Interviews with some/any parents, employers etc. who may be involved with the development of the scheme in individual schools. These will probably be conducted as and when opportunities arise, e.g. at liaison meetings, parents evenings.

2. A study of pupils

This study will focus on a sample of 20 pupils (max.), six of whom will be studied intensively. In relation to this sample, data will be gathered from:

a) Observation and audio-recording of occasions when pupils and teachers/tutors are involved in the process of discussion and recording of achievements;

b) Observation of related counselling sessions, tutor periods or lessons;

c) Follow-up interviews with teachers/tutors and pupils after a) and b) above (conducted separately);

d) Interviews with pupils, individually or in groups, at critical points in their school careers e.g. at the time of 3rd year option choices;

e) Self-concept or self-esteem inventories administered to pupils at the beginning and towards the end of the pilot project (if feasible);

f) Analysis of Documents of Record resulting from the recording process;

g) A postal questionnaire to parents. (Some interviews if possible, especially with parents of the six pupils selected for intensive study);

h) Interviews with any employers, FE/HE college tutors, YTS agents, community workers etc. who receive summative records from pupils in the sample during the life of the pilot scheme.

NOTES

(i) It is hoped that information from ALL the above sources will be collected for each of the six pupils selected for intensive study. Information about the remaining 14 pupils in the sample is likely to be less comprehensive.

(ii) It will almost certainly be necessay to secure the agreement of parents to the involvement of their children in the research. The researchers wish to discuss this with headteachers at the earliest opportunity.

Reports

The researchers hope to produce the following reports of their case study research:

a) Interim case studies of individual schools (Summer 87);

b) An interim report of across-school issues (Summer 87);

c) A final case studies report, incorporating all case studies of individual schools with some LEA (and Consortium) context (Summer 88);

d) A final across-school issues report (Summer 88).

October '85.

Funded by the Department of Education and Science

 **PILOT RECORDS OF ACHIEVEMENT IN SCHOOLS EVALUATION**

THE OPEN UNIVERSITY, SCHOOL OF EDUCATION, WALTON HALL, MILTON KEYNES, MK7 6AA
Telephone (0908) 653753 (Direct line)

UNIVERSITY OF BRISTOL, SCHOOL OF EDUCATION, 22 BERKELEY SQUARE, BRISTOL, BS8 1HP
Telephone (0272) 24161 (Ext M371)

Please reply to: EG/D4

ETHICAL GUIDELINES

Introduction

These guidelines are intended to clarify the role of the national evaluation and spell out the procedures that the evaluation team will adopt to protect individuals whilst respecting the public right to information about the pilot schemes. The guidelines are set as answers to questions that participants might wish to ask.

What is the purpose of the national evaluation?

1. The national evaluation is required to evaluate the progress and results of the Records of Achievement Pilot Schemes on behalf of the National Steering Committee (RANSC), which will be responsible for drawing up draft national guidelines. The evaluation team will pay particular attention to the objectives and issues identified in the DES Statement of Policy, but will also be sensitive to other issues and concerns that emerge during the study.

2. The evaluators will respond to requests for feedback from the pilot schemes and individual schools that participate in the evaluation study, in so far as these requests are compatible with 1. (above).

3. It is also anticipated that the general progress of the pilot schemes, and the evaluation, will be of interest to the wider educational community. We are developing plans for dissemination.

Who are the evaluation team?

The team comprises a group of researchers from the Open University School of Education and Bristol University School of Education. Their names, university base and involvement are as follows:

Dr. Patricia Broadfoot, Bristol, Co-Director, part-time;
Phil Clift, OU, part-time;
Mary James, OU, Deputy Director and Research Fellow, full-time;
Bob McCormick, OU, part-time;
Sue McMeeking, OU, Research Assistant, full-time;
Professor Desmond Nuttall, OU, Co-Director, part-time;
Dr. Barry Stierer, Bristol, Research Associate, full-time.

How will the evaluators seek access to information?

1. The evaluators will seek reasonable access to work and personnel of participating LEAs and schools, including pupils. Access to parents and employers will be sought where appropriate.

2. The evaluators will treat all relevant interviews, meetings, oral and written exchanges with participants (including DES sponsors) as 'on the record', unless specifically asked to treat them as confidential.

3. Where practicable, and providing the permission of participants has been sought, oral data will be tape-recorded and transcribed. In the case of interviews, copies of transcriptions will be returned to participants who will be given a period of fourteen days in which to amend them, if they so wish.

4. When information is sought from pupils, or when pupils are observed, the school will be invited to decide whether it is appropriate to seek the permission of parents.

238

5. The evaluators will seek reasonable access to relevant documents of pilot schemes and schools. However, they will not examine or copy files, correspondence, or other internal documents without explicit permission.

How will the data be stored?

1. Where feasible, data will be entered into micro computers for data-base and textual analysis.

2. Participants will be given access to any identifiable personal data held on computer files, in line with the Data Protection Act 1984.

3. All data, including data on disc, will be kept in archives at the Open University and Bristol University. Only members of the evaluation team will have access to these. They, and the project secretaries, will undertake to treat all data as confidential until release has been agreed with participants.

4. Data of a personal nature e.g. pupil records, will not be kept longer than is necessary for the purpose of the study.

How will the information be reported?

1. Evaluation reports can be expected to contain three elements: description, interpretation and judgement. Criteria for judgement will be made explicit.

2. Consortia and LEAs will be identified, but case study schools will be given fictitious names. Individuals will be referred to by role descriptions or pseudonyms. While this does not guarantee anonymity it reduces the likelihood that individual schools, teachers and pupils will be identified.

3. Information to be used in general summaries, which involve no specific detail about individuals or groups, will require no special clearance.

4. Permission will be sought for the reporting of statements (quotations etc.) that might identify individuals. They will also be invited to comment on the accuracy, relevance, and fairness of any report concerning them. A period of twenty one days will be allowed for participants to respond to reports in this way.

5. The evaluators will seek to improve reports in the light of such comments. However, participants will have the right to make a short written response to any report concerning them if they disagree with its contents.

August 1985.

Funded by the Department of Education and Science

SUMMARY OF RoA CRITERIA AND AREAS FOR DISCUSSION IN DES STATEMENT OF POLICY

Key:
Ordinary type – criteria that RoAs are expected to fulfil
Italic type – areas where further consideration is invited

PURPOSES
RoAs are intended to:
- provide recognition of experience and achievement
- increase motivation and personal development
- provide feedback on suitability of curriculum, teaching and organisation
- provide a summative document of record (para. 11)

KEY CHARACTERISTICS
RoAs should:
- cover the whole secondary phase
- involve all teachers in contact with pupils
- provide opportunities for appropriate teacher-pupil discussion (para. 12)
- supply positive statements only in the summative document (para. 13)

TARGET GROUP
RoAs to involve:
- all pupils in secondary phase
  (Need to investigate systems in special needs area and FE/tertiary) (paras. 14 & 15)

CONTENT
RoAs should:
- provide comprehensive coverage of whole educational programme, including extra-curricular activities
  (Need to consider further the inclusion of out-of-school activities?) (para. 16)
- include both personal and academic/skills components (para. 17)

a. Personal achievements and qualities
This component to be:
- based on teacher-pupil discussion
- diagnostic of both pupil and school strengths and weaknesses (para. 19)

- include factual list of achievements/experiences
- give contextual explanation where necessary (para. 20)

(Need to consider place for pupil accounts or teacher assessments. DES statement suggests these should be optional but if teacher assessments of personal qualities are included they should:
- *be related to observable in-school experience*
- *be positive*
- *include concrete examples*
- *be in sentence form (not grades or tick-boxes)).* (para. 21)

b. Examination results
These should:
- be included in summative document (at some point)
- with reference to explanation of coverage and grades (para. 23)

c. Other educational attainment
In relation to this component:
- internal assessment may be appropriate in formative records (less so in summative if replaced by public examinations) (para. 26)

- for non-examination pupils other skills or subject assessments (graded assessments?) could be appropriate but more credible if externally validated or if internal systems are externally accredited. (para. 27)

- all assessments should be criterion–referenced
- all assessments should contribute positively to motivation etc. (para. 28)

(Need to consider further: role of developments in graded/staged assessment systems; *(para. 29)*
role of cross-curricular assessments). *(para. 28)*

RECORDING PROCESSES/COMPILATION

RoA processes should:
- begin with summary at entry to secondary phase
- continue throughout secondary phase   (para. 35)

- involve pupils   (para. 36)

- involve all teachers in contact with a pupil, though one may need to co-ordinate   (para. 37)

(Further consideration needed of: relationship to reports to parents;   *(para. 38)*
                              storage, retrieval and transfer).   *(para. 39)*

FORMAT OF SUMMATIVE DOCUMENT

No decisions to date but national guidelines expected to specify/advise on:
- *length*
- *main components and order of appearance*
- *treatment of personal qualities*
- *presentation of evidence*   *(para. 32)*

VALIDATION AND ACCREDITATION

Several levels mentioned or implied but needing further investigation:
- *teacher-validation through the production of evidence*
- *teacher cross-validation (triangulation)*   *(para. 37)*

- *authentication of pupils' personal accounts by outsiders*   *(para. 33)*

- *external validation of assessments (graded tests or examinations)*
- *accreditation of internal assessment systems*   *(para. 27)*

OWNERSHIP

- summative documents to be the property of pupils
- master copy to be retained by the school though permission needed for
  wider distribution.   (para. 40)

(N.B. Nothing said about ownership of formative records)

RESOURCES

Resource needs to be determined:
- *INSET*
- *capitation*
- *time*   *(para. 41)*

NOTE ON THE NATURE AND STATUS OF THE INTERIM ACROSS-SITE REPORT (1987)

In our introduction to the interim across-site report we indicated that we decided to structure the report in terms of mapping of themes, illustrations from practice and issues raised. Before we finalised its content we returned the draft report to schools and local co-ordinators for clearance. We invited their responses in relation to the factual accuracy of descriptions, the fairness of interpretations and the relevance of the issues raised. We then attempted to improve the report in the light of their comments before making it generally available in its present form. Some responses however were of a general nature regarding the overall style and format of the report, derived, no doubt, from different expectations of a report such as this. These raised a number of issues about the reporting of case study which deserve attention. This appendix therefore offers a rationale and explanation of the reporting approach we chose to adopt.

Of central importance was our conscious decision to make the report issues-focused. We judged that our audiences wanted information about problematic areas in which policy decisions would need to be made. We would not therefore wish to deny that our report is structured in such a way as to indicate the issues which we feel need to be resolved. However we adopted the widest possible definition of who the relevant decision-makers are because our evidence supported our view that in developing and implementing records of achievement decisions have to be taken at all levels in the education system i.e. national, local, school and classroom. We envisaged that different groups would have different perceptions about which issues most concerned them so we made no attempt to define the level at which particular decisions would need to be taken. We regarded such definitions, in themselves, as a proper focus for policy deliberation. We saw no real difficulty in addressing several audiences by means of one report and we continue to hope that the national steering committee (RANSC), pilot project directors, and school staff, within and outside the pilot schemes, will find in the report much that is useful to their particular discussions.

By adopting the issues approach our intention was to open up, rather than close down, the forum and substance of debate. We recognise that some of the issues we mention are rhetorical or perhaps even unresolvable but we chose to retain them in our account to underline the complexity of the whole initiative. We also recognise that a focus on issues tends to draw attention to problems and may create a sense of discouragement. We would not wish to given an impression of 'gloom and doom' so we have tried to counter the inevitable tendency of issues-focussed reports towards deficiency models by selecting many illustrations which describe creative solutions to problems. In other words, although we sought to reflect the breadth of experience we believe readers will discover in our accounts much that can be regarded as 'good practice'.

We acknowledge that by identifying issues we engaged in interpretation of our evidence. We would argue that even in apparently descriptive accounts an act of interpretation is inherent in the simple selection and organisation of data. In an across-site study of this kind, however, interpretation becomes the central organising principle. We do not regard this as inconsistent with the role of an evaluation. We are aware that in the literature of evaluation theory there continues to be debate about the extent to which evaluators should themselves engage in analysis and judgement. On the one hand, there are those who advocate descriptive 'portrayal', whilst on the other hand there are those who feel that evaluators have a responsibility to make judgements based on their own independent criteria. There are also many evaluators, amongst whom we would count ourselves, who take a position somewhere between these two poles. The everyday, commonsense understanding of evaluation assumes some ascription of worth and in our 'ethical guidelines', which we agreed with participants at the start of our work, we stated that we expected to include description, interpretation and judgement in our reports. Although readers might argue that some implicit judgements have crept into this report, we have tried to be even-handed in the way we have presented issues. In all cases the issues we raise are grounded in the evidence, which we present in our illustrations. Occasionally we have 'gone beyond the data' but in these instances the issues we raise are logical extrapolations from the data. In this sense they are still firmly grounded in our evidence. Of course, we have not wished to pre-empt the judgements of readers on the issues, although we have made judgements about which issues are relevant. We would defend our right to make such judgements and, furthermore, we would wish to reserve our right to make judgements of a more substantive kind in our final report, if by then we felt we had good reason. We do not believe this gives us disproportionate influence over the national steering committee because members of RANSC are able to examine our reports in the light evidence presented in pilot scheme directors' reports, local evaluators' reports, with their own observations on scheme visits, and with the observations of HMI.

We are also more than willing to admit that there are many things which we might have done but which we have left undone, mostly because of the limitations on the time and person-power available to us, but also because we are aware of limitations on the quantity of evidence our audiences will have time to digest. Our decision to make the first substantial report, based on our case studies, an across-site issues report has meant that we have skipped a stage that is usual procedure in case study reporting. Specifically, we have not made available interim case study reports of individual institutions as we had originally planned. Faced with a virtually impossible task i.e. preparing 23 whole school reports in a very short space of time, we made a chiefly pragmatic decision to go straight to a general issues report. We judged that most of our readers would find this the most useful form of presentation. Even case study schools, who might be expected to have the greatest interest in seeing a detailed report of their particular project, had expressed a desire to have more information about other approaches whilst they were still very much at the development stage. We saw this form of reporting as a way of giving something back to schools and schemes in return for the information they have given us.

However, this decision has meant that, at this interim stage, we cannot begin to demonstrate our grasp of the holistic qualitities of developments in individual contexts. By focusing on issues rather than on schools and schemes we had to disregard to a significant degree the way in which individual schools and schemes were portrayed in the report as a whole. Our overriding criteria for selecting evidence was to give some indication of the range of progress in the particular area whilst minimising duplication. We acknowledge the importance of the whole school context (indeed this was our rationale for investing our main effort in case study work) and in our illustrations we have endeavoured to convey as much of the context as is necessary for an understanding of the issues raised. We shall be preparing individual site reports of each of the 23 institutions we have studied, although we now think it unlikely that we will produce more than two or three exhaustive case study reports in our final report. These will enable us to explore aspects that are not easily incorporated into across-site reports, notably, the interaction in particular contexts of all the various elements we have separately identified. They will also enable us to give more recognition to development over time which is, of course, an important feature of progress in all schools. In other words, we hope to compensate for some of the reductionism that is an unavoidable feature of any across-site report.

As we hope readers will recognise, we have tried to be assiduous in our avoidance of simplistic and spurious comparisons. As far as possible we have avoided even the use of quasi-quantitative language (some, many, few etc.) by presenting discrete examples of practice. Thus we have pursued only qualitative comparisons. We hope the arguments for qualitative inquiry are now well enough known and understood for us not to have to defend our approach against charges of anecdotal reporting. Suffice it to say that we have been careful to check the observations we make against our evidence. The fact that we have returned accounts to schools for clearance has been a further means of cross-ckecking our perceptions. Throughout we have tried to give recognition to the indivisibility and multi-dimensional nature of many influences in particular school contexts whilst still gaining purchase on key issues.

Finally, we realise that there is some anxiety within pilot schemes that the illustrations we have selected will be taken to be representative of schemes as a whole. We can only reiterate that there is no justification for this kind of generalisation from school to scheme. Our case study schools were not selected for their potential to typify a scheme approach but for their potential to illuminate development and implementation issues in a variety of very different contexts across all pilot schemes: special schools, selective schools, urban schools, church schools etc. It would therefore be unfair for any one school to be taken as the paradigm example of the RoA initiative in its locality even though it may have been the only school we had the resources to study in the LEA. We would remind readers that the interim meta-evaluation, which follows and is intended to complement the across-site report, gives more detail in some areas and of scheme-side issues.